MW01287478

GHOSTED

Some other titles in the Bloomsbury Sigma series

Sex on Earth by Jules Howard
Breaking the Chains of Gravity by Amy Shira Teitel
Spirals in Time by Helen Scales
A is for Arsenic by Kathryn Harkup
Suspicious Minds by Rob Brotherton
Herding Hemingway's Cats by Kat Arney
Soccermatics by David Sumpter
Wonders Beyond Numbers by Johnny Ball
Science and the City by Laurie Winkless
Built on Bones by Brenna Hassett
The Planet Factory by Elizabeth Tasker
Catching Stardust by Natalie Starkey
Nodding Off by Alice Gregory
Turned On by Kate Devlin
Clearing the Air by Tim Smedley
Sway by Pragya Agarwal
Kindred by Rebecca Wragg Sykes
First Light by Emma Chapman
Overloaded by Ginny Smith
Beasts Before Us by Elsa Panciroli
Our Biggest Experiment by Alice Bell
Wonderdog by Jules Howard
Into the Groove by Jonathan Scott
Warming Up by Madeleine Orr
The Long History of the Future by Nicole Kobie
Please Find Attached by Laura Mucha
The Great Auk by Tim Birkhead
Six Minutes to Winter by Mark Lynas
V is for Venom by Kathryn Harkup
Sink or Swim by Susannah Fisher

GHOSTED

A Social History of Ghost Hunting, and Why We Keep Looking

Alice Vernon

BLOOMSBURY SIGMA
LONDON • OXFORD • NEW YORK • NEW DELHI • SYDNEY

BLOOMSBURY SIGMA
Bloomsbury Publishing Plc
50 Bedford Square, London, WC1B 3DP, UK
Bloomsbury Publishing Ireland Limited,
29 Earlsfort Terrace, Dublin 2, D02 AY28, Ireland

BLOOMSBURY, BLOOMSBURY SIGMA and the Bloomsbury Sigma logo are trademarks of Bloomsbury Publishing Plc

First published in the United Kingdom, 2025

A catalogue record for this book is available from the British Library

ISBN: HB: 978-1-39941-8-706; eBook: 978-1-39941-8-690

2 4 6 8 10 9 7 5 3 1

Typeset in Bembo Std by Deanta Global Publishing Services, Chennai, India
Printed and bound in Great Britain by CPI Group (UK) Ltd, Croydon CR0 4YY

Contents

Introduction 1

Chapter 1: Enter the Séance Room 13

Chapter 2: Spirit Committees 39

Chapter 3: Haunted Houses 63

Chapter 4: The Ghost in the Stereoscope 91

Chapter 5: Summerland 117

Chapter 6: The Ghosts of War 141

Chapter 7: The Ghost Laboratory 165

Chapter 8: The Poltergeist Next Door 191

Chapter 9: Haunted Objects, Haunted People 217

Chapter 10: Who You Gonna Call? 241

Conclusion 269

Acknowledgements 279

References 280

Index 289

Introduction

I don't believe in ghosts. And yet.

And yet, when I watch a horror film or read a scary story, the shadows in the dark of my bedroom at night take on twisting, moving forms. Whenever someone tells me about an experience they can't explain – a passing figure who seems to vanish, something interpreted as a sign from a recently deceased relative, or strange noises in the night – my stomach fizzles with a strange sense of thrill. If I know a place is allegedly haunted, my eyes will dart to the corners and the tucked-away places just in case there's something there. And sometimes, usually as I'm falling asleep, a vacuous horror overwhelms me: death can't really be final, can it?

I don't believe in ghosts, but I want to.

In recent years, there has been a boom in ghost hunting. Local towns offer spooky nighttime tours, heritage sites are hired out for paranormal investigations, and YouTube is now awash with hour-long videos of groups recreating the techniques of their favourite ghost-hunting reality television programmes. Radio podcasts about listeners' ghostly experiences gain cult followings, and sticky-floored pubs offer evenings with spirit mediums. Psychic hotlines are a multimillion-dollar business in the US, with mediums often charging upwards of $100 for an individual reading.[1] In 2005, the Gallup Organization conducted a survey into paranormal beliefs among 1,000 Americans and found that 73 per cent believed in at least one kind of paranormal phenomenon, with 32 per cent believing in the existence of ghosts.[2]

We seem to be almost naturally disposed to be freaked out by the mention of ghosts. We can be sceptical, scoff and say it's all in our heads, a hallucination, and that ghosts don't exist, but even the toughest resolve can weaken in a dark house when the floorboard unexpectedly creaks. Death is the ultimate unknown in human experience, and the unknown scares us; in the face of such a void, our imaginations naturally conjure up things to fill it despite our fiercest sense of rationality.

Modern ghost hunters, as we'll see, are exploring ways in which small aberrations in our environments contribute to a feeling of unease. In 2009, a team of psychologists led by Professor Chris French created a 'haunted' room based on the theory that fluctuations in electromagnetic fields (EMF) can disorientate the human brain and cause us to experience unsettling sensations and even hallucinate.[3] In a bare, circular room, made dim and cool to replicate a spooky environment, EMF measurements were produced in certain areas. Participants were asked to walk around the room and, on a floor plan, indicate places where they experienced any unusual anomalies. The team found that nearly 80 per cent of 79 participants felt dizzy and disorientated at points where the spiking electromagnetic fields were located. They noted, moreover, that had the participants been told the room was haunted, the result would likely have been much higher.

It seems that we can be sceptical until we're faced with something uncanny. Ghost lore is so rich – we could all describe at length a stereotypical ghost: pale, see-through, making lights and televisions flicker, knocking objects off tables, thumping around the house at night – that we're all sitting on a wealth of expectation just waiting to be triggered by an unexpected power outage or hearsay about someone else's spooky experience. We'll see again and again in this book how many scientists and sceptics had their minds forever changed by what they experienced while ghost hunting in haunted houses and while sitting at a séance table. This is why we look for proof, I think; it's not wholly to debunk paranormal experiences as so many claim their investigations to be, but to have our minds changed.

Our history of ghosts is long and ever-evolving. The first recorded mention of a spirit features on a Babylonian tablet from the fourth century BC. As soon as we began to experience the creeping feeling of unease surrounding notions of death and the possible return of a lingering soul, we began to investigate. One of the first stories of ghost hunting comes from a letter written by the Roman lawyer and philosopher, Pliny the Younger, born in AD 61. He relates the story of a house in Athens which had been empty and for sale for some time due to its reputation for being haunted. At that time, ghosts and revenants were respected and revered, thought to seek vengeance or deliver ominous messages. But many of the Greco-Roman philosophers shrugged off religion and superstition in favour of rationality and analytical thinking, and were a precursor to some of the investigative and sceptical minds we'll meet throughout this book. Despite the overwhelming trend for fearful belief in ghosts, there were still those who did not subscribe to the idea of human spirits returning to haunt the Earth, and one of these was the philosopher Athenodorus. He was described as a *curiositatum omnium exploratory*, or an explorer of strange things.[4] Hearing the local rumour of the haunted house, Athenodorus acquired it for himself.

Not allowing himself to be taken in by the rumours and gossip, or for the house's reputation to feed into a sense of expectation, Athenodorus settled down one evening to work on his writing. Sure enough, however, he started to hear noises. He ignored it as much as he could, but the knocks and bangs and metallic clinks were fraying his nerves. The noises had begun elsewhere in the house, but were slowly getting louder, closer. When he heard the eerie sounds outside the chamber in which he was writing, he couldn't take it any longer and turned around. A ghost stood in the doorway, beckoning him to follow. The spectral man's hands were chained, and Athenodorus realised that this was the cause of the metallic noises he had been hearing.

Athenodorus steeled his resolve, turned back to his work and carried on writing. But behind him, the clinking of chains came closer and closer. Still, Athenodorus ignored the ghost. Interestingly, the ghost in this tale is described not as a fearful harbinger of doom but as a minor annoyance. It came right up to Athenodorus as he worked, rattling its glowing chains incessantly in the philosopher's ear until, finally, he caved and put his writing aside. He stood up and followed the ghost, who now eagerly drifted back through the house. It led Athenodorus to the courtyard outside, and then vanished. The following day, Athenodorus had the courtyard dug up, and discovered human remains complete with rusted chains. He removed the fetters and reinterred the bones to a more appropriate site, and never heard the ghost's rattling chains again.

This brief but fascinating case shows that despite our cultural attitudes to ghosts and the afterlife, whether we are firm believers or staunch sceptics, nothing consumes our curiosity quite like experiencing something we cannot explain.

But Athenodorus would have to wait for many hundreds of years before his method of sceptical inquiry became a leading approach to the way we investigate ghosts. Indeed, in Europe during the Middle Ages, we didn't particularly investigate ghosts at all. We didn't need to; they were part of the fabric of religion, something that was understood at the time to be intensely, terrifyingly real. The existence of ghosts was never questioned apart from finding out what they wanted from the living.

Much of ghost lore in medieval society boiled down to a dead person's ability to rest, quite literally, in peace. Funerals and burials were crucial; according to Jean-Claude Schmitt, any death that was violent or involved a religious taboo such as murder, suicide, a woman's death in childbirth or a stillborn baby, or in cases where the body had been washed away or lost, was likely to produce a revenant that would cast a 'blemish' on the family.[5] In other cases, the spirits of the dead remained to give a message, point to their murderer, or implore their relatives to seek justice for whatever mistreatment they had received in life. In Christian tradition, memorial rituals, performed regularly and particularly on anniversaries relevant to the dead person, were important ways of dispelling ghosts through seeking to cut their ties with the living and help them pass on from a lingering purgatory into a permanent heaven.

This was how ghosts remained in the public consciousness; they hadn't moved on because of stereotypical unfinished business, and it was up to the living to understand what they wanted and to act to bring them peace. Ghosts were then rather forgotten amid the rise of Protestantism, a branch of Christianity that fiercely objected to the purgatorial state as a notion and therefore rejected the very existence of ghosts. In terms of the paranormal, witchcraft and the occult was the main focus of the public although, as we'll see, the line between ghosts and demons became increasingly blurred in the early-modern era.

As Catholicism became pitted against Puritanism, exorcism became a way to demonstrate the continuation of Jesus' actions in Gadarenes, showing that there was still work to be done and that only Catholics had the inherited power to deliver victims from the power of possessing spirits. The often public performance of investigating and exorcising possessed people was a tangible miracle.

One of the most famous cases occurred in 1565, when 16-year-old Nicole Obry allegedly became possessed by the ghost of her grandfather, Joachim, at Vervins in north-east France. At first, his presence within Nicole was fairly benign. He gave her messages, asking for the family to pray for his soul at Mass and to make pilgrimages on his behalf. As Sarah Ferber describes, such instances of dead relatives appearing before or possessing members of their living family to assign tasks was not uncommon in most Christian sects, and was often seen as a way 'to have the living help them resolve their spiritual debt and speed their passage through purgatory.'[6]

The family, however, didn't do as old Joachim asked, and the spirit turned on his granddaughter, causing her to lose her sight, hearing and control over her own limbs. It was at this point that the Obry clan wondered if it really *was* Nicole's grandfather after all, and began to seek the help of the local clergy. The case was graver than expected, and Nicole was clearly very unwell, so the clergy enlisted the help of a Dominican father, Pierre de la Motte.

Father La Motte began his investigation by trying to discover the true identity of the spirit afflicting Nicole, through asking a series of questions in Latin. Ferber notes that this was a 'crucial step in the process of establishing possession by demons, as a key sign of possession was the ability to understand languages unknown to the sufferer.'[7] Next, he threw holy water in Nicole's face, causing her to recoil – revealing that it was, indeed, a demon with an ingrained revulsion for holy objects that inhabited the girl's body. The demon, disguise ruined by the holy water, revealed that he was Beelzebub.

Now that the demon had been brought to light and named, he didn't hold back on tormenting Nicole. She started to exhibit the same behaviour of the Gadarenes demoniac exorcised by Jesus: she writhed and raved, spat blasphemous gibberish, and threw her body around with such uncanny force that three adult men couldn't hold her down.

Ghosts were at best messengers, and at worst phantom pollutants who needed to be avoided or cast out through rituals. In France, for example, a tile would be removed from the roof of a house where someone had died in order to ensure their soul wasn't trapped within the walls. Candles were lit, prayers were given, priests called in – all to stop a spirit from lingering around their deathbed. No one wanted to see a ghost.

As we'll see, however, this attitude towards hauntings would dramatically change in the nineteenth century.

This is not a book about ghosts. It's a book about the living, and how, over the last 200 years, we have searched for proof of the existence of ghosts. Whether spirits are residual energy replaying like a broken vinyl record or intelligent beings residing in a utopian afterlife, or whether they simply don't exist at all, we have used ever-evolving techniques to attempt to capture evidence. This book tells the story of these ghost hunters, and why our fascination with the paranormal is as timeless as the ghosts we try to find.

While Athenodorus was investigating spirits 2,000 years ago, there's a reason why this book is focused only on the last 200 years. Cases such as Athenodorus' haunted house are few and far between, with the majority of ghost stories being ones of religious reverence and superstition. Ghosts were an accepted part of most people's understanding of the world. They existed to provide warnings, to seek revenge, or to beg for a proper burial; they delivered a message, and then they disappeared. There was no need to investigate something everyone accepted as truth.

The Renaissance and its scientific revolution, however, led to a greater split between spirituality and rational ideas of life and the universe. Religion as a whole, while still prominent, didn't dictate people's lives in quite the same way. Natural philosophers were endlessly developing their understanding of the human body and fatal diseases, and superstition over illnesses and sudden death diminished. Ghosts and their spooky messages were beginning to appear less and less to their relatives, their roles largely redundant.

In the nineteenth century, our relationship with the possibility of an afterlife shifted dramatically, and ushered in the age of ghost hunting as we know it today.

We'll begin our journey here, examining how two American sisters began the worldwide craze of séances and how, for the first time, we actually started to *communicate* with ghosts rather than simply listen to their warnings. Mediums who claimed to be better in tune with the spirit world rose to celebrity status, earning impressive wealth through acting as the conduit for ghosts to speak with their loved ones. But as the new religion of Spiritualism developed, so too did an equal and opposite movement of sceptics rise up to investigate and challenge the notion of chatty ghosts. For each renowned medium, there was also a notorious critic whose methods of exposure were unorthodox, ingenious and sometimes downright violent. As we'll see, though, their scepticism could be shaky – and many a scientist ferociously opposed to Spiritualism gradually became converted.

In the wake of mediums and scientists equally being burned by their paranormal dabblings, societies were set up to organise and combine efforts, providing guidelines and peer support as well as to legitimise the practice of investigating ghostly phenomena. These societies and associations produced some of the most infamous clairvoyants and gave scientists the space and means to expose fraudulent mediums.

In the twentieth century, the fierce debate around ghosts shifted as the world grieved for the young men brutally killed in the First World War. During these years of catastrophic loss, Spiritualism was welcomed in the homes of the bereaved as a way for families to speak to their sons, brothers and husbands who had died so far away from them. More than ever, the public sought evidence of ghosts in order to feel as though their loved ones – often in what should have been the prime of their life – were living on in a utopian Spiritualist heaven. It's in these moments in the history of ghost hunting that scepticism becomes a tricky stance to take. On the one hand, mediums were exploiting the grief of millions for a tidy profit; but, as we'll see, the comfort such messages from the afterlife brought to families was something that was difficult to dismiss.

As the world began to heal from its losses, ghost hunting changed again. Individuals broke from psychical research societies to investigate haunted houses and phantoms through self-funded experiments, though not without controversy, stoked by competitors and rivals. The mid-twentieth century was a time of media frenzies surrounding these ghost hunters, whose cases made for excellent sensational journalism. Some of the most popular stories in the UK were those of poltergeists – loud and violent spirits who tore apart homes from the inside.

But while the poltergeists wrecked homes in the UK, demons seemed to haunt adolescents in the US. Stirred by the Satanic panic of the latter decades of the twentieth century, teenagers and children were feared to be increasingly under the control of demonic spirits who infiltrated their minds through spirit-communication tools such as the Ouija board. Ghost hunters in the US combined their techniques with those of the exorcist, and spirits began not just to haunt locations, but people and objects, too.

At the turn of the millennium, however, there was a dramatic increase in the popularity of paranormal investigations. Fuelled by the success of reality television programmes such as *Most Haunted* and *Ghost Adventures*, amateur ghost hunters packed their bags with devices and gadgets and set about recreating the chaotic investigations they witnessed on their screens. In more recent years, the majority of these ghost hunters create their own programmes for YouTube or as a live-stream – both to capture evidence but also, of course, to gain notoriety among the community of paranormal investigators.

This book will follow the timeline from the Fox sisters' first séances to the modern paranormal investigators who use technology to capture evidence. Along the way, I will talk to parapsychologists, YouTube ghost hunters and members of psychical research societies in order to find out why we continue to look for proof of spirits after so many hundreds of years of investigations. I'll examine my own scepticism in relation to my research, to try to figure out for myself why I'm drawn to the search, too.

If I could talk to a ghost, it'd be my grandmother – to have a conversation with her as a woman rather than as an adolescent.

She passed away nearly ten years ago, when I was in my early twenties. My grandfather had died a couple of years earlier, and while she managed on her own for a while she gradually became too frail to live without the support of round-the-clock care. Her mind was sharp until the end, but in the last few months of her life she'd taken to sitting up in bed at night, not sleeping, and just holding my grandfather's wristwatch in her hand to watch the time pass. The last time I saw her, during a visit to her care home, she had caught a nasty cough that rattled in her lungs. She clung onto me extra tightly as I left, and I kissed the wispy white hair on top of her head – something I'd never done before but suddenly felt compelled to do. I still think about this a lot; neither of us said anything, but somehow, something inside of us both knew that this was our last goodbye. I think, among other things that I've been recently discovering, I inherited her intense discomfort at having people fuss over her. I feel like we connected on a new level in that final moment, both agreeing, quietly, that we wouldn't make this worse for each other by admitting what we both knew was coming. Our gestures said everything we needed to say. She was taken to hospital later that afternoon, and died a few days later.

I'm now in my thirties, firmly in adulthood; my hair has become wavy over the last few years like Grandma's, my cheeks and mouth a little more defined and serious. I resemble her more and more as I get older; I look at old photographs of her and I see my own dark eyes. When I first started obsessively watching old Hollywood movies, my mum told me how much Grandma loved them, too. I'd never known this about her, and now I won't ever get the chance to talk to her about my favourite films and actors.

Even without the concept of ghosts, people haunt us. But it's not just people – it's places, things. When Grandma went into a care home, we had to clear out and sell the bungalow she had lived in since she married my grandpa nearly 70 years before. It was a small bungalow on a corner plot, so the garden wrapped around it and my sisters and I spent many afternoons doing endless circuits. Inside, the bungalow hadn't changed much since the mid-century. My grandpa had been in the Merchant Navy during the war, and loved buying knick-knacks from his travels, so it was full of lovely little 1940s-style souvenirs from around the world. The walls outside were textured and painted cream, with a green front door. It's strange what stays with us; I miss that green door. I miss it to the point that when I managed to get on the property ladder a couple of years ago, buying myself a little flat in a small, friendly complex, I subconsciously decorated it in a way that recreates the feel of that bungalow. The furniture is of a dark, mid-century style and, wherever I could, there are little green accents that remind me of the bright green door. My favourite of my grandpa's souvenirs, which I used to rush to look at every time I visited as a child, is a little wooden weather-house he brought back from Innsbruck in Austria. A tiny man and woman come out of respective doors depending on fine or wet weather – I don't think it really works but I love it anyway, and it sits prominently on a shelf in my living room.

If ghosts are echoes of the past, then I've created my own haunted house. This is what I think ghosts are; I think we create them ourselves from the things about a person – their voice, their scent, the objects that surrounded them – that linger most in our minds. I don't think there is anything after death besides what the living continue to cling to. But perhaps writing this book will change my opinion.

I don't believe in ghosts, but I'm going to look for them.

CHAPTER ONE

Enter the Séance Room

Near the British Library in London, there is a haunted apartment. It was once owned in the first half of the twentieth century by ghost hunter Elliott O'Donnell, who didn't stay there long as there was, he said, something 'indescribably evil in the atmosphere.'[1] Soon after vacating the premises, he learned the reason for its unpleasant aura: it had once been the site of a séance – a ritual to contact the dead.

The apartment's previous residents had been a widow, Mrs Dale, and her beloved son Arthur, whom Mrs Dale smothered with affection. But when Arthur died suddenly at the age of 21, Mrs Dale's grief was all-consuming. Unable to cope with the loss of her son, she employed a spirit medium and held a séance.

With the gas lights in the apartment turned low, Mrs Dale, two friends and the medium sat around a table, willing the spirit of Arthur to come through, to return to his bereft mother.

They waited.

For over half an hour, they waited. And then the table began to twitch. Strange creaks and knocks were heard in answer to the party's questions. A spirit had arrived, the table told them, but it wasn't Arthur. It bore, instead, a message from Mrs Dale's son to say he 'regretted his inability' to appear before them.

Mrs Dale wouldn't accept such a message. Arthur *had* to come. Why – she demanded of the medium – why wouldn't he come? She insisted the medium, whose protests to leave the boy alone fell flat, make him appear before her. Against her better judgement, the medium relented, and they began again to wait, the clock ticking slowly closer to midnight.

Suddenly, Mrs Dale exclaimed that Arthur had arrived – she could smell his cologne. The others gathered around the table sniffed, but could only detect the scent of something nasty and decaying.

Again, Mrs Dale insisted that the medium bring Arthur forward. He was so close now, after all; she could smell him around her. The medium 'gave a convulsive shudder' and fell silent. Then Mrs Dale appeared to have her wish granted, but not, perhaps, in the way she had expected. She gasped and, before fainting, cried, 'Oh, go away, for mercy's sake, take it away!'

O'Donnell tells us that whatever Mrs Dale had seen, it 'would haunt her to her dying day' and she begged those gathered around the séance table 'never to mention Arthur Dale to her again'.

She moved out of the apartment shortly after, but the foul presence remained. The séance had brought *something* out, and it wouldn't go back.

The séance was the Holy Communion of what became known as the Spiritualist movement; practised millions of times in public halls, theatres and private homes, it was the foundation on which Spiritualism was built. In 1848, two young women, Kate and Margaret Fox, were asleep in their home on a farm in Hydesville, New York. They were awoken by strange sounds: bumps, creaks, knocks and scratching that were unlike the usual noises of bats in the rafters or of a house settling against the cold of the night. They knew that their farmhouse was haunted by the ghost of a man named Mr Splitfoot, who had been murdered there, and were sure that this was his restless spirit. Attempting to communicate with him, they found when they asked him questions, he would respond with knocks. The sisters went on to create a code, with a certain number of knocks to correspond to 'yes' and 'no' answers, and found – in a rather dramatic departure from previous ghost lore – that Mr Splitfoot was self-aware, and could speak intelligently to them.

Remarkably, they could use his knocks on the walls of their home as a form of communication with the dead.

This wasn't by any means innovative. In China, during the Ming dynasty (1368–1644), *fuji* séances used a planchette which spirits could move to write messages in sand or ash. In London, moreover, a ghost inhabiting a building in Cock Lane in 1762 knocked intelligently when asked questions. Somehow, though, the Fox sisters created a craze out of their conversations with the dead. People flocked to see the girls talk with the sentient ghost, and news of the astounding breakthrough to the afterlife travelled quickly. The séance format designed by the Fox sisters was found to be easily repeatable, and Spiritualism as an organised religion was born. Within a few years, swathes of young women professed themselves to be similar operators of the 'spiritual telegraph', as the séance system was first known. They were the mediums through which, at a séance table, the spirits would channel their energy and tap out messages. And their followers increased exponentially. By 1877, it was estimated that the number of Americans who aligned themselves with

Spiritualism was over 11 million, and it was not long before the phenomenon moved abroad as mediums travelled on tour to Britain, France and Australia.[2]

In the decades following Mr Splitfoot's first message, Spiritualism became a serious business, and a very profitable one. By the mid-1850s Kate Fox had attracted a patron in American manufacturer Horace Day, who was paying her an annual sum of $1,200 (£972), equivalent to around $43,000 (£34,800) today – and that was simply for the privilege of the occasional private séance.[3] Soon, there were enough followers for a sense of elitism to grow among those who undertook séances in a professional capacity, as opposed to those who did it for a laugh. Emma Hardinge Britten, one of the most prominent advocators for Spiritualism in England, sought to streamline people's efforts and instigate good practice when communicating with the dead. In the February 1868 edition of Spiritualist publication *Human Nature*, she outlines her 'Rules to be Observed for the Spirit-Circle' in order to gain the best possible results from ghosts.[4]

She was very specific about the conditions necessary for a successful séance. Those gathered around the table, somewhere between three and twelve (but preferably an even number), should be 'of opposite temperaments, as positive and negative in disposition, whether male or female; also of moral characters, pure minds, and not marked by repulsive points of either physical or mental condition'. Chronically ill people should not be allowed to participate, nor those who are interminably bland in character (or words to that effect).

The influence of mesmerism and animal magnetism comes through in Britten's rules, but we also see the beginnings of what would be one of Spiritualism's defining characteristics: its members' positioning of the movement as a scientific endeavour. She writes:

> The use growing out of the association of differing temperaments is to form a battery on the principle of

> electricity or galvanism, composed of positive and negative elements, the sum of which should be unequal. No person of a very strongly positive temperament or disposition should be present, as any such magnetic spheres emanating from the circle will overpower that of the spirits, who must always be positive to the circle in order to produce phenomena. It is not desirable to have more than two already well-developed mediums in a circle, mediums always absorbing the magnetism of the rest of the party, hence, when there are too many present, the force, being divided, cannot operate successfully with any.

There is a sense that, like a chemist preparing their laboratory, factors such as contamination, interference and calibration need to be taken into account. The room must be cool and well ventilated, the table must be wooden as it is the most effective conductor of spirits, and there is to be no strong source of light which, Britten says, produces 'excessive motion in the atmosphere, [and] disturbs the manifestations'. This latter point in her rules is a significant criticism of sceptics as they started to take aim at Spiritualism; it allows the ambiguity of semi-darkness to be maintained and, as we will see in a later chapter, prevents the use of flash photography without first duly warning the spirits.

For all the emphasis on the exacting and scientific properties of various elements within the séance room, there is still an aspect in Britten's rules that demonstrate Spiritualism's similarity to religion and communal rituals. Unless a sceptical investigator put a stop to the practice to focus purely on gathering evidence, most séances began with contemplative conversation, and a group rendition of a religious song or poem – some of which were written specifically to be used in such a situation – and Britten encouraged this. A particularly popular poem recited at the beginning of a séance was 'The Other World' (1867) by Harriet Beecher Stowe, American author of *Uncle Tom's*

Cabin and fervent Spiritualist. The poem is a kind of hymn that praises spirits in the afterlife:

> It lies around us like a cloud,
> A world we do not see;
> Yet the sweet closing of an eye
> May bring us there to be.
> Its gentle breezes fan our cheek;
> Amid our worldly cares,
> Its gentle voices whisper love,
> And mingle with our prayers.
> Sweet hearts around us throb and beat,
> Sweet helping hands are stirred,
> And palpitates the veil between
> With breathings almost heard.[5]

The poem is quite long and rambling, perfect to sing slowly while waiting for spirits to reveal themselves. It ends with the lines, 'Your joy be the reality,/Our suffering like the dream.' While this poem was written fairly early in relation to the timeline of Spiritualism and psychical research, it demonstrates a feeling that would carry the debate well into the twentieth century and beyond: that sense of yearning for some scrap of proof – even the slightest breeze – that gave the 'suffering' living hope for continued life not just for themselves, but for those they had lost. Spiritualism was, after all, borne out of grief. Around the time of the Fox sisters, there were mass outbreaks of disease such as typhoid, cholera and tuberculosis, and during the 1860s, when Spiritualism truly gripped America, families experienced widespread grief in the Civil War. As critic Molly McGarry demonstrates, medical advancements had been promising great progress in curing illnesses and healing wounds; the combination of war and epidemics 'undid the expectation [...] that medical science could reduce mortality'.[6] Spiritualism arrived precisely at a time where people faced death from multiple angles, and their blind faith in God's heaven would no longer ease their uncertainty: they needed to search for heaven themselves, while still alive, and find proof that the soul lives on after death.

Mediums who began in small, local Spiritualist circles became celebrities, often embarking on international tours to demonstrate their talents to packed audiences in theatres and concert halls. Spiritualism rapidly moved from private gatherings to declaring new proof of the afterlife on a public scale through news articles, journals, magazines and books, and through the endorsement of well-known figures. It was only to be expected, then, that an equal and opposite movement of sceptics refuting the 'evidence' rose up. But the divide between Spiritualists and their scientist critics wasn't so clear-cut, and a crisis of trust and reputation soon developed.

The most renowned early example of this was the morally dubious 'working' relationship between celebrity medium Florence Cook and esteemed chemist William Crookes.

Florence Cook was a young, impish girl of 17 who had already caused a stir among the older, competitive mediums who disliked the new kid on the circuit. Nevertheless, she enjoyed rapturous success, and by 1873 was deemed worthy of investigation by Crookes. Cook's séances demonstrated a new technique that was sweeping through Spiritualist circles. Table-tilting was already old hat, and people were dissatisfied with receiving messages from their loved ones through the creaking of wooden boards. But Cook's skills extended far beyond simply making contact with the dead – she could bring the dead back to life.

During a séance, Cook would retreat into a cabinet set up precisely so that she could sit in a trance, undisturbed by those gathered around the table. After some time, a small peep-hole door in the cabinet would open, and the eerie face of a spirit would be visible for a brief moment before the cabinet was again sealed. Soon, though, even this was not enough to maintain her popularity, and Cook instead began to materialise the spirit of a single dead girl, Katie King. This wasn't as new as people thought. Another celebrity medium, Daniel Dunglas Home, had been pushing the boundaries of spirit phenomena through magnificent feats of levitation – and had been moving the role of the medium more towards that of a conjuror than the operator of the 'spirit telegraph'. Cook, too, had been dabbling in levitation, but her new talent for materialising King brought her fame to unprecedented heights. She would summon the spirit of the dead girl, draped in voluminous folds of white fabric, who would emerge from the cabinet to greet and kiss the members of the circle.

Where Florence Cook was young, slight of build and with dark hair, Katie King was also young, slight ... and with dark hair. Suspicious. Cook was closely guarded by her parents, who admitted guests to her séances under strict invitation and prior approval (though approval could, of course, be bought). And yet somehow, they allowed a man named Mr William Volckman to attend a séance and meet King in the ghostly flesh. The séance occurred in Hackney in December 1873, with a large number of people present, including members of the aristocracy such as the Earl and Lady of Caithness, and the Count El Conde de Medina Pomár. They assembled around the room, eager to witness Cook's alleged powers.

As the fabled Katie King emerged from the séance cabinet, the assembled party were fascinated and enraptured by the strange, elfish spirit who walked among them. Mr Volckman waited like a cat among pigeons, until King moved closer to him.

He leapt out of his chair, seizing the spirit girl around her waist, trying to unbalance her and cause her to fall to the floor. What he found, of course, was that King was flesh and blood. She was, as he suspected, Florence Cook wearing a bedsheet over her head. He thought his exposure would be the end of Cook's career, and no doubt enjoyed a moment of triumph as he felt the girl's solidity in his arms. But it was merely a brief moment. Two other men in the séance dashed to their feet and pulled Mr Volckman by his neck away from

King, leaving him with a bloody nose. Another man, Mr Luxmoore, chivalrously escorted the shocked and ailing King back to the privacy of the séance cabinet.

The rest of the séance attendees, who considered themselves to be Cook's true friends and allies, wasted no time in writing a character assassination of Volckman and defence of Cook and sending it to the *Spiritualist*, entitled 'Gross Outrage at the Spirit Circle'.[7] By acting quickly they perhaps thought they'd be able to diminish trust in Volckman before he had the chance to write up his side of the story.

In the defending article, the group claimed that after King had been led back to the safety of her cabinet, she asked them to wait for five minutes. In time, the curtain opened and Florence Cook, looking sorely wounded and exhausted, reappeared before the group. The clothes she had been wearing before the séance began were still in place; not a stitch had been broken. There was no trace of the white outfit worn by Katie King, nor was there anything in Cook's pockets. But there was one change: Cook was in severe pain. She spoke of having been burned. Volckman's act of seizing King had only served to transmit a terrible injury to her medium, and now Cook was the one who suffered.

After a few weeks, Volckman was persuaded by his friends to finally publish his own report of what happened at the séance. Trying first the *Spiritualist*, his letter of protestation was rejected; it was clear whose side the editors were on. Instead, he wrote a lengthy article to another Spiritualist journal, *Medium and Daybreak*, under the heading 'My Ghost Experience: The Struggling "Ghost"'.[8] He included the unpublished letter to the *Spiritualist*, claiming that Katie King was '*no ghost*, but the medium, Miss Florence Cook, herself.' He also tried to downplay his physical assault. He said he first observed the materialised spirit of Katie King for 40 minutes, in which he behaved himself completely. Then, he took her hand in his, but purely because she had previously said that the séance sitters would be allowed to do so. It was only when her hand turned out to be suspiciously firm and warm that Volckman was possessed by the idea of seizing her around the waist – quoting Luxmoore's description of his actions as 'an outrage', but nonetheless admitting to it.

This is something we will unfortunately see time and time again in the history of ghost hunting, particularly where female mediums are involved. Certain types of investigators took the mediums' elusive, dreamlike qualities, and the vulnerable position in which they placed themselves in order to communicate with the dead, as in 'invitation' to poke, probe and, in short, harass them in a way that seemed to be considered acceptable conduct in the séance room. In giving part of herself to the spirit world, the medium also gave up much of her autonomy.

But Volckman was not, in fact, a total sceptic. Indeed, the reason why his 'experiment' involved taking Katie King's hand is that he believed that in other séances, with other materialised spirits, he had held genuine spirit hands that were translucent and ethereal, and whose grip he could never quite hold. He quoted an article in the *Quarterly Journal of Science*, in which the author described a similar experience of seizing a spirit's hand, only for it to turn to vapour. Volckman was not denying genuine spirit phenomena – far from it – he was simply exposing a single case of fraud.

The Spiritualists were ferocious in their defence of Katie King and Florence Cook and did everything they could, from lectures to books, to dismiss Volckman's evidence. They focused more on his despicable conduct, drawing attention away from what he had actually discovered, and showing instead that the man was a brute whose words couldn't be trusted. The damage, however, was done, and Volckman's seizure of Katie King had left a permanent mark on Cook's reputation. But her parents, who had been enjoying their daughter's success, weren't done yet, and were determined to find someone who would assure both Spiritualists and sceptics of Cook's genuine powers.

That someone was William Crookes, the author of the very article Volckman quoted when discussing genuine spirit hands. From 1873 to 1874, Crookes was invited to attend Cook's séances to investigate her powers through a series of scientific experiments. He was a well-respected figure in British chemistry, Fellow of the Royal Society, and discoverer of the element Thallium. Before he was 16 years old, he was already engaged at work in chemical laboratories, and published his first research paper at the age of 19. He'd had a long, successful career. He was also, however, thoroughly interested in conducting experiments with Spiritualist mediums. Prior to his investigations of Cook, he had gained notoriety by investigating Daniel Dunglas Home, and published his findings not in one of the growing number of sensationalist Spiritualist publications, but in the *Quarterly Journal of Science*. He explained that his stance was from a sense of responsibility as a scientist; strange phenomena had been reported, and so it was the 'the duty of scientific men' to examine such things which 'attract the attention of the public, in order to confirm their genuineness, or to explain, if possible, the delusions of the honest and to expose the tricks of deceivers.'[9] He invented a copper wire cage, basically a big waste-paper basket, into which Home was asked to lower an accordion, holding only one end of the instrument. With Home's left hand placed in view on the table, and his right hand holding the accordion in the cage, there was, Crookes believed, no way for Home himself to emit notes from the instrument.

The resulting phenomena were reported to 'baffle explanation'. The accordion wasn't one Home had brought with him – Crookes had bought it new, especially for the experiment – and thus there wasn't a chance that Home had tampered with it in any way.

Yet, once the séance began, the accordion started to quiver. Witnessed by everyone present, it started to emit breathy, wheezing sounds that grew longer and stronger and more defined until 'several notes were played in succession.' But Home wasn't moving. His left hand remained flat on the table, and the hand that gripped the accordion in the cage was held still.

Soon, the accordion was moving with an uncanny vitality, swinging around and around the cage, not just playing random notes but whole tunes. Crookes' assistant, a man whose observation skills were trusted entirely by the chemist, looked under the table and decided that Home couldn't be the source of the movement. Seizing Home's arm, Crookes felt along the muscles and tendons. Surely if he *were* somehow moving the accordion, Crookes would have detected the ripple beneath Home's skin. But there was nothing. The medium was utterly, remarkably still, even as the accordion was performing a merry jig around the cage.

Crookes couldn't explain or replicate the phenomena, and presented his research exactly as it was witnessed, thinking, perhaps, that his colleagues would be equally interested in Home's talents. They weren't. Instead, he was inundated with letters criticising his experiment and apparent gullibility, and worse still, was publicly mocked in newspapers and other scientific journals. Remarkably, however, Crookes was undeterred and continued to submit papers about further experiments and ruminations on Spiritualist phenomena.

'Doubt', he told his opponents, 'but do not deny; point out, by the severest criticism, what are considered fallacies in my experimental tests, and suggest more conclusive trials; but do not let us hastily call our senses lying witnesses merely because they testify against preconceptions'.

After the saga with Mr Volckmann seizing Katie King, Cook's parents believed there was no one better suited in the world of science to defend Florence's skills than William Crookes. He began his discussion of his lengthy investigations with her by explaining the matter of light in séances, noting that many of the experiments with Cook took place in the light. However, most of the episodes he described seem to be in darkness, or, at least, crucial phenomena were hidden away from the light. It is here that we see the development of Crookes' involvement with Spiritualism moving from curious scientist to full-blown believer. 'Light', he said, 'exerts an interfering action' on mediums and ghostly phenomena, particularly when the spirit is a weak one. In any case, he said, spirits often produce phenomena *as* light – as luminous fireflies buzzing around the room, as flashes in answer to a question, as glowing clouds floating above heads – and how can that be effectively observed if the surroundings aren't sufficiently dark?

And not only did Crookes' beliefs change, but so did the audience to which he presented his findings. He wrote now to Spiritualistic journals rather than scientific ones about Cook's mediumship, and in February 1874 accused her critics of bringing forward 'very few *facts*' in their case against her.

In the first séance witnessed by Crookes, he saw Florence Cook disappear into her cabinet – this time, a back drawing room separated by a curtain. Shortly afterwards, Katie King emerged, but only for a brief moment, excusing herself by saying that Florence wasn't well enough to sustain her materialised form for very long. At no point in this initial séance did Crookes see Florence and Katie at the same time, and in fact admits that in the dimness of the room there was a stark resemblance between the spirit girl and her medium, but he insisted that in the brief moment Katie appeared from behind the curtain, he could still hear Florence making the same moaning sound she had been emitting as she went into her mediumistic trance in the cabinet.

The following month, Crookes described another séance. Again, he contradicted his earlier remark that the use of light is permissible in séances, and that many he attended did take place in the light. It's clear in the instalment of March 1874 that the séances were not, in fact, well-lit at all. In this instalment he mentioned experimenting with using phosphorus oil in a small, corked, 6- or 8oz bottle, and allowing its sickly glow to help him distinguish between the faces of Katie and Florence. During the March séance, held this time in Crookes' own home with his library acting as the medium's cabinet, Katie emerged once more and, surprisingly, asked Crookes into the library to help Florence, whose head needed supporting. Sure enough, there was a woman in the library and, according to Crookes, only three seconds had passed between Katie asking Crookes to see Florence and Crookes entering the library. Crookes went up to Florence, who had slipped from her original position on the sofa and was now half-draped on the floor, and lifted her to a more comfortable and secure spot. If Crookes' estimation of three seconds was accurate, then there wouldn't have been time for Florence to change out of Katie's white robes, back into her own clothes, and flop on the sofa.

But, equally, and crucially, he didn't see the two women together. Katie vanished as soon as she had delivered her message, and wasn't in the room when Crookes went to reposition Florence. Why wouldn't she have stayed to supervise the way Crookes handled her medium? Perhaps such a comment was made, because Katie told Crookes to follow her into the cabinet to see the two girls together. Taking his phosphorus bottle to light the way, he entered the library once more. 'I looked round for Katie,' he writes, 'but she had disappeared. I called her, but there was no answer.'

Confused, he resumed his seat in the séance room. Katie reappeared again a moment later and told him 'she had been standing close to Miss Cook all the time'. If Crookes had any suspicions at this point, he didn't express them.

In a later séance, Crookes finally got the evidence he desperately sought: Florence and Katie, together in a room. This time, however, there was no light permitted – not even his feeble phosphorous bottle – until he had reached the place where the medium was positioned, having slumped off the sofa in a trance. Fumbling his way across the room, his hands touched Florence. The lamp he brought in with him was lit, and he turned to find Katie, at last, standing behind him and dressed in her usual white robe and wimple-like hat. Katie said nothing to him, but he was sure she was really there, and that it was really Florence on the floor. He checked each girl three times, examining them 'with steadfast scrutiny'. When Florence began to wake from her trance, Katie quickly ushered Crookes out of the cabinet again.

Crookes defended Florence almost to the point of ruining his own career. He was relentlessly hounded by critics, even well into the twentieth century when it was revealed that Crookes' opinion of Florence may have been helped by a few secret intimate exchanges. Meanwhile, Katie insisted she had to go back to the spirit world forever, and eventually Florence's talents faded into obscurity.

Nevertheless, despite numerous other controversies among Spiritualism, it continued to garner further popularity. Among Spiritualism's high-profile advocates was Sir Arthur Conan Doyle, author of the Sherlock Holmes novels*, but his belief tested his friendship with the magician Harry Houdini.

Harry Houdini took centre stage for Spiritualism's sceptics and argued ceaselessly with Doyle (although, for the most part, good-naturedly) about what each considered to be evidence in the debate. Magicians and theatrical conjurors joined ranks with scientists to denounce Spiritualist séances, seeing the practice of materialisation and ghostly levitation as a mockery of their art. At least the audience *knew* there was a trick behind the spectacles they saw on stage when they went to see Houdini and his contemporaries. While magicians came after Spiritualism from the same angle as scientists – that of exposing fraud – their stance was more of a moral one; after all, their audience members paid to be knowingly deceived. It was entertainment. Spiritualism, according to this group, used the same tricks and legerdemain

---

* There is a tremendous amount of irony here since Holmes is renowned for his powers of observation and rational conclusions.

as magicians, only mediums professed their tricks to be of ghostly origin and exploited their paying customers' grief and hope in an afterlife.

The magicians were having none of it. John Henry Anderson, self-styled professor and known as 'The Great Wizard of the North', was one of the first to declare war against Spiritualists. In his 1853 book, *A Shilling's Worth of Magic, or, Tricks to be Learnt in a Train* – a cheap, accessible publication bought at one railway station bookstand and dumped in another station's bin – he includes a lengthy tirade against mediums which he titles 'The Magic of Spirit Rapping, Writing Mediums and Table Turning &c., &c.'[10] As the rest of the book gives instructions on magic tricks, from card games to ones involving a mix of combustible chemicals (perhaps best *not* learnt in a train), it's clear from the outset exactly what Anderson means by 'magic' – legerdemain and audience manipulation. It's also clear from the outset that he really, *really* hates Spiritualism. From the opening lines he calls the movement 'the greatest trick of the age as practised by the most unscrupled of jugglers', and that table-rapping is 'the grand modern climax of all impostures'. Despite being a Scottish magician, Anderson concerns himself mostly with America, and each time he rants about Spiritualism's manic spread across the homeland of the Fox sisters he adds extra zeros to the number of believers from thousands to tens of thousands and hundreds of thousands. He sees it as a disease, a viral infection that sends rational minds foolish and delusional.

He was quick off the mark, too. Only four years had passed since Kate and Margaret took their table-rapping talents to public audiences, but it was long enough to arouse significant suspicion and, in Anderson's and many professional magicians' case, outright disgust. The Fox sisters had a lot to answer for. He describes them as a pair of silly yet scheming girls who came up with the knocking technique to 'terrify their parents', but on finding how successful their trick was, their 'vanity or their natural tendency to deception caused them to follow up the joke until the matter grew serious'. Serious and, indeed, extremely lucrative. Anderson claims that in just four years the Fox sisters had amassed a fortune of $500,000 (£405,000) which is equivalent to an eye-watering $20 million (£16 million) in today's money. That's a staggering amount for stage performances of séances, but, given Anderson's tendency for exaggeration as discussed above, is perhaps not entirely accurate. Nevertheless, their modest lives on the farm were changed completely, allowing them to build their own mansion and travel in comfort and style, so they did indeed make an extortionate amount of money very quickly.

Anderson made it his mission to expose the tricks of the Fox sisters and their followers. He claimed it was because Spiritualism gave people false hope, even turned them mad, but there must have been some professional offence taken, too. Spiritualism, after all, was big business – why would people pay to see tricks they *knew* to be fake, when they could instead spend their money on what were deemed genuine supernatural phenomena? Mediums were leaving magicians in their dust, so it was up to magicians to maintain their relevance through the exposure of Spiritualism.

One of Anderson's methods was particularly cunning, and played on what we've recently discussed: the celebrity spirit. While in the US, Anderson gathered some like-minded friends keen to investigate the veracity of table-rapping. They decided to attend separate séances, at the same time, and each – on the dot – was to ask for a particular spirit to grace the room with their ghostly presence. They agreed on George Washington. Anderson says that the event was done mostly as a joke, but what the group found was that George Washington duly appeared at each séance, seven in total, at the very same moment. Anderson leaves it to 'the rapping philosophers' to come to a conclusion on such a result. Nothing in the lore of Spiritualism suggests that a spirit can be omnipresent; it detaches from the body at death, but remains whole, and can only be in one place at any one time. For George Washington to be in seven different séance rooms, the conclusion was that either at least six mediums were frauds, or that the former US president was some sort of omnipresent god.

This test, as Anderson described, was a bit of fun between sceptical friends. However, he then explained a more inventive and scientific method by which Spiritualist mediums conducted their fraudulent séances. Apparently taking inspiration from a rumbled medium, Anderson created his own table set-up with which he conjured up his own spirits. It required the physics of electromagnetism, a galvanic battery, some wires and a hidden accomplice. The battery was kept in a cupboard or in another room with the fellow conspirator, with wires passed underneath the carpet or floorboards up to the medium's table. In a secret compartment in a table leg or on the underside of the table surface, a piece of metal connected to the wires was placed next to a lever with an iron plate at one side and a block of wood at the other. When the battery was connected by the accomplice – close enough to the medium to hear everything said within the room – the metal in the table became magnetic and attracted the iron plate, drawing the wooden block on the other side of the lever upwards to make a rapping sound. When the battery was disconnected again, the iron plate was released and the wooden block clunked downwards to produce a second noise.

So proud was Anderson of this invention that he made it part of his show, and challenged any medium to join him on stage and make spirits rap upon his table without the use of the galvanic battery. No one took up his offer, and so he continued to make his exposure of Spiritualism a key part of his act. Indeed, many magicians during the initial buzz of Spiritualism actually drew upon the movement in their stage performances. Much of their new material re-enacted common experiences and phenomena at séances, and then showed the audience exactly how the uncanny movement and noises were created through simple means.

Anderson died the year Harry Houdini was born in Budapest in 1874, almost as though Houdini was destined to continue Anderson's legacy of exposing fraudulent Spiritualist mediums. As Houdini's career and fame blossomed, he too set about making his own investigations into mediums and séances. They are some of the most fascinating – and trustworthy – of this period in Spiritualism's history. Chronicled in his book *A Magician Among the Spirits* (1924), he describes the various séances he and some of his friends attended, and the cunning tricks he uncovered along the way. But Houdini's scepticism didn't entirely come from his sense of professionalism; it also came from his own aching grief. Having recently lost his mother, on whom he doted, he confessed that it was largely from being in such an emotionally vulnerable position that he became so incensed by the practice of mediums. He claimed that he made pacts with a number of his friends, colleagues and relatives that whoever died first would send a spirit-message to the other.[11] Fourteen members of this group had passed away by the time he wrote his book, and for all the séances and sittings Houdini investigated over 30 years, not one person came through via a medium.

And in those 30 years, he only became less convinced of the alleged powers of mediums. As we saw with Crookes, and will continue to see throughout this book, there was a tendency for sceptical investigators – scientists especially – to become slowly convinced by Spiritualism, usually because they found in themselves a growing attraction to the medium and thus took a less objective view of the phenomena. Not so with Houdini. Indeed, Houdini criticised the previous work of scientists whose investigations became more and more morally compromised:

> There is no doubt in my mind that some of these scientists are sincere in their belief but unfortunately it is through this very sincerity that thousands become converts. The fact that they are scientists does not endow them with an especial gift

> for detecting the particular sort of fraud used by mediums, nor does it bar them from being deceived, especially when they are fortified in their belief by grief, for the various books and records of the subject are replete with deceptions practised on noted scientists who have essayed to investigate prominent mediums.

The only good, reliable investigators of Spiritualism, according to Houdini, were magicians. After all, scientists try to compare the physical and the spiritual, the real with the unreal. Magicians, on the other hand, were simply comparing tricks. A scientist may be able to record temperature differences or analyse chemical residue found in a medium's cabinet, but they're not trained to detect sleights of hand, camouflaged prosthetics and planted accomplices.

But the line between Spiritualism and stage-magic was blurry, and some of the most prominent mediums of the time professed, usually when they'd been exposed, that they never claimed truly to contact the spirits, they were only performing conjurors. This was precisely the case with the dashingly handsome Davenport brothers from Buffalo, New York. Ira and William enjoyed tremendous success off the back of the Fox sisters' new trend for table-tilting. The brothers would be seated and tied up beneath an array of bells and musical instruments inside a large cabinet. Once the cabinet doors were closed, discordant music would begin to play; it was thought that the spirits the Davenports had summoned were operating the instruments, as it was impossible for the brothers to move against their restraints.

Due to their fame, they also attracted swathes of sceptics, and were known to place bear traps around the dark séance room to stop any investigators from leaving their seat and poking around the cabinet. But they were exposed multiple times, and were the favourite targets of magicians who could easily work out how the phenomena were produced. Later, long after William had died of tuberculosis aged 36 and Ira had retired, Ira confessed all to Houdini, and 'positively disclaimed Spiritualistic power [...] saying repeatedly that he and his brother never claimed to be mediums or pretended their work to be Spiritualistic'. Nevertheless, they did tell their audience that they could communicate with the dead inside their spirit cabinet, and whether this was merely a theatrical performance or not, the brothers would have been well aware of the moral complexities of claiming to contact ghosts. At a time where numerous mediums *did* insist that they were talking with the dead, it would be foolish to 'pretend' to do the same and not expect the audience to take them seriously.

By this point, however, even high-profile exposures and the downfall of mediums such as the Fox sisters (who, later in life, admitted their conversations with Mr Splitfoot had been fake) and Daniel Dunglas Home couldn't put a pin in Spiritualism. It had grown far beyond its humble farmhouse roots and mediums had established a number of new techniques for contacting spirits, each with more cunning trickery than the last. Houdini systematically took down a vast array of these phenomena. The classic knocks on wood were produced by 'slipping a knee up and down against a table leg' or strapping a wooden block 'under the skirt' for a sharper sound. In the dark of the séance room, the medium Eusapia Palladino would, while holding hands around the table, subtly move until her neighbours either side held each other's hands instead of her own, leaving her free to perform phenomena. To write spooky messages on slate boards, Henry Slade made a ring in which a fragment of pencil lead could be attached, allowing him to scribble surreptitiously on the surface. For every strange occurrence, it seemed, there was a cunning trick behind it.

Spiritualism had grown too vast and too varied to be brought down by the exposé of one celebrity magician. It carried on despite Houdini's best efforts to change the mind of the general public. But, as we'll see, it was only after

Houdini's death in 1926 that Spiritualism began to be investigated more seriously, and the true ghost hunters emerged to debunk the idea of life after death.

When Professor Anderson, The Great Wizard of the North, wrote his exposure of Spiritualism, he briefly mentioned the way the unconscious mind can play a key role in séance phenomena. This was in 1853, well before Sigmund Freud and his contemporaries developed their theories of psychoanalysis, hypnotism and the way we can be driven to action without knowing our true intentions. Anderson described how, when an ordinary person sits down at a séance table, the conditions and context are such that 'each party wills the table to go round, though he does not will his hand to move it'. This contradiction is a battle that happens in the unconscious mind, and the unconscious mind responds by moving the muscles in the person's fingers, thereby causing the table to creep in a certain direction. Interestingly, Anderson makes a prediction, regretting that 'there is no deeper philosophy than this – that no new physical force has been discovered – that the mind of an individual cannot be made to exert a greater influence on inanimate substances'.

Over a century later, this is precisely how one group of investigators who were all members of Toronto's Society of Psychical Research, a local branch of an investigative movement, approached séance phenomena. In 1972, the society selected a group of eight people to conduct a series of séances that are now known as the 'Philip Experiment', overseen by Dr A. R. G. Owen. The group was selected to be a representative sample of the general public – among them was an accountant, engineer and a number of 'housewives'.[12] Their mission was to conduct séances as a group, without, notably, a medium's influence, to summon the spirit of a seventeenth-century English aristocrat named Philip.

Only, Philip wasn't real; the Toronto society asked the group to invent a spirit, to see if spiritual phenomena would still occur when knowingly trying to summon a person who never existed. Philip's ghost was a product of fiction – intentional, deliberate fiction whose backstory was fleshed out over the course of a year, although the group's combined knowledge of history was sometimes patchy and meant that Philip's life was filled with historical inaccuracies. Before his death, he had been in a political marriage with a beautiful woman called Dorothea, but the marriage was a loveless one. The group become delightfully melodramatic, and decided to make Philip commit adultery with a raven-haired gypsy called Margo whom he stumbled across on his land. He brought Margo back as a secret mistress to his home, Diddington Manor. But, plot twist: Dorothea found out about Margo and, to save her reputation and enact revenge on Philip, had the girl executed for witchcraft. Philip was so horrified at this turn of events that he threw himself off the battlements of Diddington Manor (which, incidentally, is a real place *without* battlements). Poor Philip.

The group spent initial sessions discussing Philip, learning his story and wondering how he might react to certain historical events or figures such as Oliver Cromwell. One member, an artist, drew a simple portrait of Philip to help them keep a visual image in their minds. They sat around a table, and quietly meditated together, keeping their thoughts and conversation centred on the character they had created. Over time, they began to feel as though he had really existed, and sometimes the group had to remind themselves of the experiment: the whole point was to bear in mind that he was a product of their collective imagination.

But nothing happened. The group was disappointed. They had spent a year trying to conjure up the spirit of their fictitious character, and it seemed as though that time had been wasted. Perhaps it was an impossible experiment, after all. They were close to giving up entirely.

The team had one last idea. Previously, the sessions had been purely about meditation, almost like a group grievance therapy session where everyone discussed the kind of man they knew Philip to be. They decided to do something different, to change the dynamic of the sessions. Instead, they would style the room and the meetings like a Victorian séance. By the 1970s, séances were deeply unfashionable, and had been largely left behind in the wider cultural history of ghost hunting. But the group felt there might just be something in it. They followed rules such as those lined out by Emma Hardinge Britten, and began the séance with songs and prayer, light conversation, and were positioned around the table with their palms flat on the surface. It was important, according to the report, for the group to accept that this new approach was a tried-and-tested method for the production of paranormal phenomena.

The séances began, but still nothing happened. Then, on the third or fourth session, the group experienced something new.

Underneath their hands, they felt a rapid vibration shudder through the table. Philip was among them. The vibrations turned into more defined raps, three for yes and one for no, just like a Victorian séance. The table was made of plastic, rather than a spiritually conductive wood as Hardinge Britten suggests, but that didn't seem to matter to Philip. He made sure they knew he was present, rapping out three times when asked if it was really him. But no one suspected that any member of the group was fraudulently producing the raps – why would they? They all wanted to see the experiment produce genuine results.

The phenomena quickly grew in intensity. The table started to glide unnaturally across the carpeted floor (which, when they tried to recreate the movement, only moved in a juddering sort of way) and to rear up on two legs, even one leg, and the group seemed to have trouble keeping hold of it sometimes. The report describes how the table 'slid away from their hands so quickly that it was difficult to keep up with it. It would be impossible for anyone to be pushing it without the rest of the group being aware of it.'

In this new, exciting phase of the experiment, the group received results beyond their wildest imaginings. It wasn't only raps they heard, but rumbling percussive peals. In a more sinister moment, the group asked him about his villainous wife Dorothea and were met with animalistic scratching sounds beneath the table. If the group placed their hands on the table, they could feel the way energy seemed to change in the room as Philip was summoned. The table would 'creak and groan, as if something was attempting to wrench it apart.' It was as though pressure was building through the table as Philip was forcing himself from the afterlife to the material world, from fiction to reality, and suddenly the table would almost explode with vigorous raps and movement as he made himself known.

The experiment was set up to see if the group could collectively hallucinate Philip, but these results are so much more peculiar. They show how the role of expectation is potentially the key to many ghostly experiences, particularly ones situated around a séance table.

Perhaps the only ghosts around us are the ones we create in our own minds.

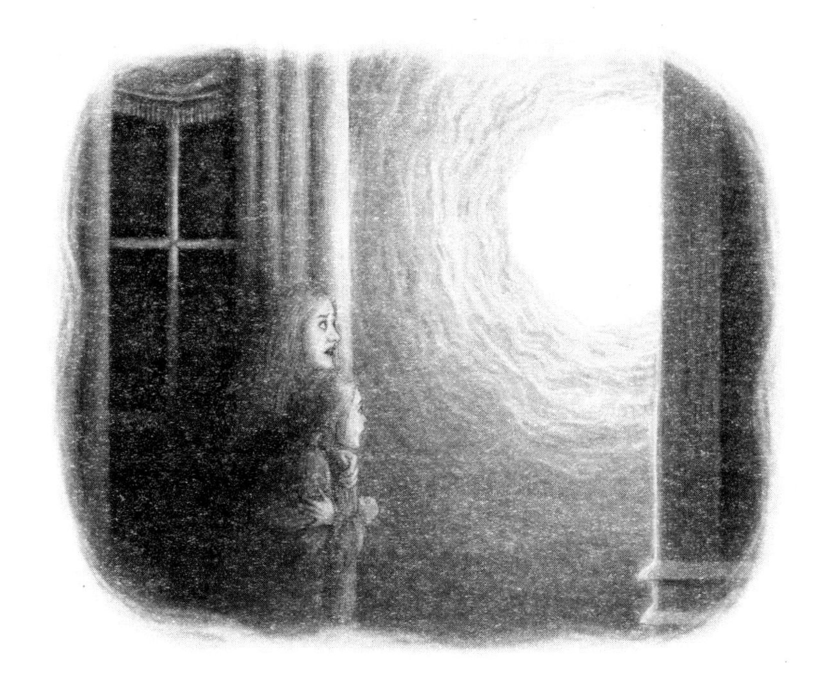

CHAPTER TWO

Spirit Committees

As Spiritualism became further scrutinised by science, the forms and methods which scientists were using to test mediums were becoming increasingly uncontrolled. To add to that, scientists' reputations were often at stake, and conducting their tests as independent researchers opened them up to fierce public criticism. It seemed as though everyone was interested in Spiritualism in some capacity, but few in organised and 'respectable' scientific circles had the gall to discuss and investigate it publicly. William Crookes, writing in 1874, demonstrates a sense of isolation in his efforts to investigate Spiritualism and the medium Daniel Dunglas Home. 'I still feel', he says, 'that it would be better were such a committee of known men to be formed, who would meet Mr Home in a fair and unbiased manner, and I would gladly assist in its formation; but the difficulties in the way are great'.[1] He was announcing a common feeling among the growing number of scientists interested in paranormal phenomena. They couldn't work alone and retain their credibility; alone, they could easily be targeted by critics, their reputation ruined because they conducted their own tests without the support of their peers.

This was the foundation for the formation of ghost-hunting societies. The earliest of these was founded in Cambridge, England as the Cambridge Association for Spiritual Inquiry, affectionately known as The Ghost Society, in 1851 by the Reverend B. F. Westcott and the Bishop of Durham – interestingly, men of the clergy rather than of science. The early days of The Ghost Society have unfortunately passed into obscurity, as noted by its late president Peter Underwood, and records only really begin with the organisation's second iteration, the Ghost Club, which is believed to have been established in London in the early 1860s. For the large part, the club at this time was secretive and private, more of an exclusive social club for members to 'relate their strange experiences' in an informal, friendly, like-minded atmosphere.[2] They had no public journal, nor did they advertise for new members – any new member had to be suggested by the president and each member had to agree to allow him in. Members were rather illustrious men, including Sir Arthur Conan Doyle, high-ranking military men, the journalist W. T. Stead, poet W. B. Yeats, and Sir William Crookes. In a rather masonic tradition, members would address each other as 'Brother Ghost', and it was only in 1929 that the first woman was allowed to become a member.

The focus of the club at this point was not investigation, but rather story-telling. Many ghost stories at the time, by writers such as M. R. James, were framed as an intellectual fireside chat in which the sceptical academic related events that continued to haunt them, and in the same vein, each meeting of the Ghost Club operated on the basis that members took turns relating their own inexplicable experiences. But, as Underwood demonstrates, there were only so many times members could listen to the story about a phantom toilet-flusher, or the time someone probably mistook a sheep for a ghost, before interest fizzled out and fewer and fewer people attended the organisation's meetings and dinners.

Besides, another society – one that actively sought out evidence of ghosts – was being formed, and was attracting far more interest and far more members.

In 1866, three friends sat down together to discuss their common interest in alleged paranormal manifestations. They were Frederic Myers, Edmund Gurney and Frank Podmore, and they approached Spiritualism from an inherently sceptical point of view. They decided to write a two-volume book, *Phantasms of the Living*, in which they explored, and tried to explain via theories of illusion, medicine and mental illness, various experiences that seemed to have no earthly cause. They were interested in 'correcting or neutralising individual fancies or exaggerations [...] leaving as little as possible to the unchecked idiosyncrasy of any single thinker.'[3] Despite their scepticism, there was a sense throughout the book that Myers, Gurney and Podmore *did* believe in some sort of paranormal force, but mostly in respect to psychic power and what they called 'thought transference'.

Then, in 1881, a conversation between the journalist Edmund Rogers and physicist Sir William F. Barrett, who both wanted to create a more organised society for the research of unexplained phenomena, led to a conference at the British National Association of Spiritualists. A committee was formed at the conference which would go on to establish the Society for Psychical Research (SPR), and included Rogers and Barrett as well as Myers and Gurney. In February 1882, they elected philosopher Henry Sidgwick – whose wife was the fellow investigator and eminent physicist Eleanor Sidgwick, sister of Prime Minister Arthur Balfour – as their first president. We'll return to Eleanor later.

The society outlined their objectives as organising and streamlining investigations into the paranormal, which until that point had largely been done by individuals. By working as a group, they could pool resources, cover more ground and keep more meticulous records of investigations. Researchers were also, for the first time, offered a modicum of protection by conducting their work for a society backed by some of the most important thinkers and public faces of the time. To that end, the SPR created a series of specific subcommittees that investigated thought-reading, mesmerism, Reichenbach's experiments (which related to the notion of the human aura), apparitions and haunted houses, physical phenomena and a literary committee. The aim, as illustrated in the first volume of the *Proceedings of the Society for Psychical Research* (1883), was to 'approach these various problems without prejudice or prepossession of any kind, and in the same spirit of exact and unimpassioned inquiry which has enabled Science to solve so many problems, once not less obscure nor less hotly debated.'[4] The subscription fee cost two guineas for fully fledged members, and one guinea for associates – around three days' pay for a skilled tradesman, and worth around £80–£90 in today's money. Their first general meeting, held on 17 July, 1882, was opened with an address by Henry Sidgwick, who spoke of the Society as a combined effort, and as a means to gather evidence of phenomena on an industrial scale. He didn't refute the work of individual investigators, but argued singular pieces of evidence by lone researchers were not enough to convince the general public of the reality or fraud of certain paranormal experiences.

The Society wasted no time in getting to work, and the first volume of the *Proceedings* contains detailed reports of investigations from all committees, as well as individual lectures, addresses, presentations and anecdotal reports. As a group, they grew in number very quickly, and attracted the interest of several well-known scientists and thinkers, such as Sir Arthur Conan Doyle (again), Nobel Prize winner John

William Strutt, and Sigmund Freud. The painter and patron of the pre-Raphaelite brothers, John Ruskin, was also a keen member – mostly because he sought painting advice from long-dead masters.[5] While many in the committee were scientists, the membership as a whole covered a large (albeit very middle class) number of professions, from writers to physicians to members of the nobility.

Two years after he helped to form the Society for Psychical Research, William Barrett took a trip to the US, the birthplace of much of the phenomena the Society sought to investigate. While there, he met with several like-minded men who were scientists interested in psychical phenomena, and encouraged them to set up the American Society for Psychical Research. Among its early members was Alexander Graham Bell, inventor of the telephone, and psychologist William James, brother of writer Henry James whose fiction – especially *The Turn of the Screw* (1898) – explored the idea of the dead returning to haunt the living. The James brothers had been raised among esoteric beliefs; their father, Henry James Sr., was a renowned Swedenborgian theologian, and household conversations often turned to ideas of spiritual existence, metaphysics and psychical phenomena. But where Henry chose to explore his interest through terrifying ghost stories, William turned his attention to investigating real-life cases of hauntings, séances and telepathy.

Some of the best work conducted by the British Society for Psychical Research in its early years was done by Henry Sidgwick's wife, Eleanor, who was a physicist assisting John William Strutt and an advocator for the higher education of women. Eleanor had been involved in the scientific investigation of Spiritualism well before the establishment of the SPR. As we saw in the previous chapter, eminent scientists – most often men – began to study ghostly phenomena from a sceptical point of view before soon becoming sucked in by the eerie rituals of the séance. Time and time again, the rational scientist was swayed by the careful manipulation of the medium. Eleanor Sidgwick's beliefs, however, were more difficult to challenge.

She often made mediums tremble before her sceptical scrutiny. Frequently assisted by fellow SPR member and researcher Alice Johnson, she used critical thinking to pull apart the claims of famous mediums and Spiritualists such as William H. Mumler, who professed to be able to photograph ghosts on command (more on him in Chapter 4).

In SPR meetings, Eleanor delivered scathing papers on her latest research and investigations. During a general meeting in May 1886, she gave a lecture entitled 'Results of a Personal Investigation into the Physical Phenomena of Spiritualism', and it is the work that perhaps best demonstrates her approach to ghost hunting. She discussed her initial dabblings with séances, experiences that most people had in one way or another by the 1870s, but it was only when she participated in a séance with Kate Fox – of the infamous Fox sisters – that she seemed to be particularly spurred on to investigate and expose the trickery of mediums. Her writing oozes with quick wit and absolutely brutal takedowns. Describing the numerous private séances she attended held by Kate Fox, she quipped there were 'no conclusive results, except with the discovery that she or her "spirits" are willing to claim, as Spiritualistic phenomena, accidental occurrences quite unconnected with her presence.'[6] But by this point, the Fox sisters were old news, and in 1874 Eleanor instead set her sights on Annie Fairlamb Mellon and her colleague, C. E. Wood, who were highly renowned mediums based in Newcastle. Annie Mellon was rather infamous for materialising the spirit of a girl called Cissie, in a performance not too dissimilar from that of Florence Cook and her spirit, Katie King. Indeed, it was because of William Crookes' experiments with Cook, the results of which had just been published a few months earlier, that Eleanor and her associates decided to see if they could conduct a similar investigation (but without, hopefully, falling in love with both the medium and her materialised ghost).

Together with her husband Henry, as well as Frederick Myers and Edmund Gurney, Eleanor began to test the powers of both mediums. In one elaborate experiment, they restrained Mellon and Wood to marble pillars with leather straps, secured with wax seals, tape and padlocks requiring letter combinations to open. The phenomena produced was unremarkable. Eleanor described Mellon's materialised spirit as a 'doll or mere drapery', and when raps were heard she wondered if the medium 'might have brought something with her to make these with'. After one materialisation séance, Eleanor asked if she could search Annie Mellon, who 'sharply and decidedly declined'. The tape and seals around the mediums were often disturbed and marked with boot polish, and knots were found to be differently tied and much tighter after a séance than Eleanor had made them in the first place, despite the SPR members being present in the room for the duration of the séance. 'At any rate', Eleanor wrote, 'the indications of deception were palpable and sufficient, and we were not surprised to hear a few months later that a more aggressive investigator had violated the rules of the séance, and captured Miss Wood impersonating the "spirit".'

At this point in the history of Spiritualist phenomena, mediums were using a new technique called 'slate writing'. On top of the séance table, in the darkness, an ordinary piece of slate would be placed and a question posed to the visiting spirit. After a while, the lights would be raised again and a ghostly message in eerie, trembling handwriting would have been written on the slate. It was quicker than getting spirits to rap out the alphabet for every word, and brought the ghost that much closer to the living.

Eleanor Sidgwick was, as always, unimpressed. In 1876, she attended a series of 10 sittings with the American medium Dr Henry Slade, who was on tour in England. He was renowned for producing slate writing, and, indeed, during these séances his spirits wrote an abundance of messages. Eleanor watched him with her usual hawk-like scrutiny, noting how the messages in other languages seemed straight out of a tourist's book of conversational phrases. She watched him so carefully that she noticed instantly how often the slate was out of view. 'I am inclined to think,' she wrote in her SPR report, 'that no one ever sees the slate quite continuously from the moment they see it blank on both sides to the moment they see writing on it.' Slade came up with various cunning ruses to ensure opportunities to swap slates with pre-prepared messages or quickly scribble something on the surface with a tiny fragment of pencil. At one séance, held completely in near-darkness, he declared he wasn't feeling well, laying the foundations for a later moment when he apparently took a turn for the worse and ushered Eleanor and her associates out of the room. When they came back, there were, of course, messages on the slate.

While a scientist, Eleanor seemed to sympathise particularly with magicians' critical view of Spiritualism, and aimed to use scientific methods to expose mediums as mere conjurors. She enlisted her friend and amateur magician, whom she called Mr A., to recreate some of the séances she experienced with Slade. She noted how 'it was a shock to me to find how easily Mr A. could deceive me'; the phenomena produced were exactly the same, only Mr A. showed Eleanor exactly how a few sleights of hand could make the messages appear to have been written by a ghostly hand.

The UK had its own well-known slate-writing medium in William Eglinton, and Eleanor managed to get him to agree to three sittings with her. He already had a tarnished reputation after his materialised spirit, 'Abdullah', was, like Katie King, seized by a sceptic in 1876. Unlike Volckman and the King case, however, the proof of the fraud was pretty absolute, with psychical researcher Thomas Colley pulling the fake beard from Abdullah's face. Clearly aware of the risk, and of Eleanor's record of exposing mediums under her steely gaze, he thought it better not to produce anything at all than to be caught by her. In her lecture for the SPR, she described how the séances were 'perfect blanks, and he is almost the only medium out of 18 or 20 I have sat with, with whom I have witnessed no phenomena at all.'

Eleanor continued her work with slate-writing mediums for several years, and the culmination of her damning reports and lectures irreversibly damaged the veracity of the technique. While handwritten messages from spirits continued to be popular, the use of slates declined, and a number of Spiritualist SPR members resigned from the Society as a result of her investigations.

Eleanor's work was never done, and we'll meet her again during our look at spirit photography. There was always a sense of cynical disappointment in her investigations; she didn't have the rage of Houdini, nor the awestruck incredulity of Crookes. She was just very bemused by Spiritualism, both by the tricks of its mediums and the gullibility of its followers. As she scathingly wrote towards the end of her 1886 lecture, 'One cannot investigate Spiritualism for long without learning that some people quite sincerely think things inexplicable which they could do themselves if they tried.' The more I research the history of ghost hunting, the more I see that she was absolutely correct.

Another of the most prominent early members of the Society was Sir Oliver Lodge, physicist and innovator of early radio technology. During the latter decades of the nineteenth century, Lodge was a hugely influential scientist who had a keen private interest in Spiritualism which developed at around the same time as the Society for Psychical Research was forming. He was an early member of the Ghost Club, and between 1901 and 1903 was the president of the SPR. He was particularly interested in séances and mediumship (we'll see the true personal extent of this in a later chapter), and attended sittings with both obscure and well-known mediums.

In the 1880s, while Lodge's interest in Spiritualism and the paranormal was blossoming, he took on a young Welsh research assistant, Benjamin Davies. Thousands of letters were exchanged between the two during Davies' employment – and friendship – with Lodge, many of which are kept with Davies' papers in the National Library of Wales in Aberystwyth. Over a series of several long afternoons, I trawled through these letters. While the majority of them regard Lodge's work at the University of Liverpool and a developing laboratory at the University of Birmingham which Davies was overseeing, mentions of Spiritualism become increasingly frequent as time went on. Lodge touched upon the subject only occasionally at first, but then started to ask Davies to attend and report back on sittings in his stead. Davies, who was clearly in awe of his employer and mentor, dutifully went to these séances, and gradually grew more and more interested in Spiritualism himself.

As Davies became increasingly involved, he started to conduct his own tests and sittings with local mediums in Birmingham. What is amusing about his papers is the way in which they perfectly encapsulate the attempt to see psychical research as a natural extension of scientific enquiry. The letters from Lodge, and Davies' notebooks, are full of physics terminology and messy fountain-pen sketches of circuitry, radio waves and equipment. But their treatment of Spiritualism and mediums is almost indistinguishable from their discussions of the laboratory, and you have to pay close attention not to miss a letter about Spiritualism amid the notes about physics. In one of Davies' notebooks, for example, he appears to have been developing some sort of invention to test whether the phenomena produced by mediums is genuine.[7] Using the same rough sketches and diagrams as he used for laboratory equipment and experiments, Davies came up with what was essentially an electric-shock device. A medium, holding two conductors, completed a circuit of electrical current. Should they remove their hands from either of the metal rods in order to engage in fraudulent activity, the yelp from the ensuing shock would be enough to alert those present that the phenomena were in fact tricks. It doesn't appear that Davies ever used his device in any of the sittings he attended, but as we'll see, even if he did it wouldn't be the worst thing an investigator has ever done to a medium.

In the UK currently there are what's known as the 'Big 3' ghost-hunting societies. We've looked at the SPR and the Ghost Club – the two oldest of these societies. But in 1981, a new society was formed by members of the SPR who decided to split from this institution to create their own: the Association for the Scientific Study of Anomalous Phenomena (ASSAP). The SPR was, and still is, mostly concerned with ghosts, apparitions, the potential of an afterlife and telepathy. The founding members of the ASSAP, including *Fortean Times* editor Bob Rickard, established the society to include investigations of UFOs and cryptozoology (involving the search for legendary animals such as Bigfoot and the Mothman) – though their main interest and activity still centres on ghost and poltergeist phenomena.

The ASSAP has investigators across the country who are at the ready to respond to reports of suspected paranormal activity. They keep meticulous case notes of their enquiries, and now have quite the extensive archive. Bill Eyre, their archivist, sent me a few choice reports that showcase the kind of investigations the ASSAP conduct.

In January 2007, for example, investigators were called to a house in Castlefield, Manchester. Its occupant, a woman in her sixties, had asked the ASSAP to attend after a series of peculiar events had taken place. At first, she saw coloured lights dance spontaneously in her living room, but things soon seemed to escalate. A black shadow floated about the room until she told it to go away, ornaments moved, a strange scent of garden flowers and honeysuckle appeared in random patches about the house. The television turned itself off and on and flicked through channels without her even touching the remote control, the cat bed shifted position and the cat itself – ordinarily a chilled-out little thing – kept being spooked into a frenzy that made it dart about the house in panic.

The investigators arrived. What's interesting about this report is the lack of extravagant equipment or dramatic ghost hunting. The majority of the report focuses on gathering as many facts about the woman's experience as possible, cross-referenced with questions about her background and lifestyle. There were questions in the report about her well-being, medical history, eyesight and general enjoyment of life. The interview proceeded, focusing on the phenomena that prompted her to call the ASSAP in the first place. Through this, we find out that she had a history of involvement with the paranormal, having lived in an allegedly haunted house during the 1970s. She also, apparently, visited mediums regularly from this point onwards. We'll see in a later chapter that while mediums were certainly at the height of their popularity in the nineteenth century, they seemed to have a particular resurgence in the latter half of the twentieth century – when the woman in this case often sought their advice and help. In fact, she had contacted a medium to see what information she could glean about the shadowy presence who was scaring her cat before reaching out to the ASSAP. The medium had told her that it wasn't one particular ghost, but several spirits of the woman's relatives come to check on her (who clearly weren't fans of cats and argued over the television channels).

Some of the more intriguing facets of the report are the questions where the woman was asked how she felt during and after the phenomena. 'Gobsmacked,' she said, and then, when it was over, she simply felt 'amused'. These, to me, seem like the strangely moderate responses of someone who was acutely aware of and open to paranormal occurrences. The investigators didn't draw a conclusion, a little suspicious, perhaps, of the fact that the woman was so keenly interested in the occult in the first place.

An earlier case is far more complex. In 1996, the ASSAP was alerted by a chain of hearsay to an allegedly haunted school in Donwick (a fictional name to anonymise the location). The caretaker's family, living in a flat within the school building, had for a whole year been suffering increasingly frightening and violent disturbances that seemed to have no explanation. The caretaker, his wife and their adult son told only a few people, but eventually the reports made their way to the ASSAP and it was invited to investigate the phenomena.

There had, it seems, always been something quite eerie about the old Church of England school. The caretaker and his family had previously encountered strange electrical disturbances, occasional apparitions, and were often frightened by the severely disturbed sleep of the son. But in mid-1995, some minor work was undertaken at the school: a chimney was dismantled and furniture was moved around. This disruption seemed to trigger more frequent and threatening phenomena. Interestingly, there does often appear to be a link between construction work or general chaos within a building and reports of hauntings, as though knocking down a wall or exposing original floorboards angers the resident ghosts.

Following the alterations to the chimney, things changed quite quickly and dramatically in the school, particularly for the caretaker and his family. As with the Castlefield investigation, the ASSAP investigators, which included archivist Bill Eyre himself, conducted an interview with the caretaker's son. While all three of the family experienced frightening phenomena, it was the son who seemed to encounter the brunt of the haunting, and had the most details to share. Aged 22, he was short-sighted and asthmatic, but other than experiencing the occasional feeling of being watched, and the odd precognitive dream, he didn't have any interest in or prior knowledge of the paranormal. This stands in contrast with the woman in the Castlefield haunting, who seemed particularly interested in mediums, Spiritualism and occult phenomena. Throughout the report, there are instances described that seem to be classic signs of disordered sleep – he experienced sleep paralysis often, and frequently talked in his sleep in an eerie, childlike voice. This can, perhaps, be dismissed as purely natural, but these incidents make up only a small part of his, and his parents', experiences. Some episodes of his sleep paralysis were accompanied by equally inexplicable phenomena.

What's also fascinating about the case is the lengths to which the family went in order to keep it under wraps. They didn't, after all, call the ASSAP directly. As it was a Church of England school, they were concerned that the 'authorities' (presumably the school council) wouldn't take kindly to them announcing that the building was haunted. The family requested that there be no publicity, and that the case was kept well away from the tabloids. The caretaker's son was worried that his father would lose his job if the reports became local news. On the other hand, when asked about how the hauntings made them feel, they replied that they were very frightened. While the woman in Castlefield may, perhaps, have welcomed the attention, it's clear from the second case that this family were genuinely terrified by what was happening to them.

One of the creepiest aspects of the case is not what happened to the caretaker and his family, but to their dog. They had a little West Highland terrier, small dogs known for their gleaming white fur. The son relates a particularly terrifying night in which he had sleep paralysis while the dog was sleeping at the end of the bed. As he lay there, unable to move, the dog suddenly yelped and bolted around the room as though spooked. But it's not as though the son startled her since he couldn't move a muscle. When the paralysis finally lifted, the son went to comfort the dog and found that her ear was bleeding without any visible wound or infection.

Indeed when the ASSAP investigators began to conduct their interview of the caretaker's son, they were shocked to see that the dog's ear started to spontaneously bleed again. The son's grandfather had suffered notorious ear problems in his later life, leading the son to believe he could be one of the spirits haunting the family. There may, of course, have been a perfectly good explanation for this mysterious affliction (indeed, it's not mentioned whether the family took the dog to the vet), but it occurred amid a flurry of other strange phenomena that took place during the interview. A disembodied breeze passed through them, the temperature changed dramatically, and dark shapes were seen by both the investigators and the resident family.

Another investigator was sent to explore the flat and the school building. In addition to the family's experiences, an OFSTED school inspector had previously reported feeling uneasy in the top corridor of the school (whether this made it to the final report isn't known – it would be fascinating if it did). While walking around the school, the ASSAP investigator felt an invisible pencil jabbed into his arm, followed by the sensation of a child's hand in his. Like the caretaker's son, he saw apparitions, too. The next step was to hold a series of vigils within the flat, and there was no shortage of inexplicable phenomena. Once again, the dog's ear bled without any apparent cause, its white fur staining bright red.

The interview, the investigation and the vigils all led the ASSAP to the conclusion that this was, in fact, paranormal phenomena. Individually, the occurrences could be attributed to sleep paralysis or faulty wiring in the school, but collectively the picture was far more curious and sinister. The caretaker's family, so disturbed by the incidents and unable to bear it any longer, bought a cheap second-hand caravan to keep in the school grounds, and took to living in that instead of in their appointed flat. The investigators decided that the frequency, complexity and variety of phenomena suggested that there was more than one ghost at work here – a whole class roster still lingered, unleashed when work began on the chimney. There was a particular male ghost that was responsible for the more violent attacks on the caretaker's wife and son, but there were also an untold number of other spirits in the school building which were, at worst, just a little bit mischievous. The investigators also noted the presence of the son's grandfather, whom it was believed had come to try to defend his relatives from some of the more troublesome ghosts.

Throughout the investigation, the son's experiences were a great deal more prominent than anyone else's. He saw more, felt more, and seemed to be the most affected and the most frightened by whatever haunted the school and caretaker's flat. There was no doubt in the ASSAP team's mind that the phenomena were genuine, and that the son was particularly open and sensitive to psychic disturbances. The vigils and interviews complete, the next step was to try to actually *do* something about the haunted school. Interestingly, the interventions and recommendations were of a wholly Spiritualist nature; while the investigation itself used scientific methods, the conclusions drawn from the data were that this was a genuine haunting, and thus only spiritual solutions could be provided. They brought in a medium and his wife to conduct a 'spirit rescue' – a practice which aims to help ghosts 'pass on' and move away from the mortal plane, basically the Spiritualist equivalent of scooping the pest up in a cup and chucking it outside. Allegedly, the medium, standing in the son's bedroom, sensed the ghost of a very tall, aggressive man who had worked in the school and had stolen things from the building on more than one occasion. This man matched the description given by the son, who had seen him as an apparition, although the medium had no knowledge of this. Upon encountering the medium, the malevolent man gave him a headache and caused his eyes to water. After a bit of resistance, the medium eventually managed to encourage the man to move on and leave the school – and the caretaker's family – in peace.

Following a tour of the school, the medium returned to the son's bedroom and announced that the tall, violent male ghost had, indeed, gone. Almost a month later, the caretaker's wife told the ASSAP that the situation had considerably improved, although strange phenomena still seemed to plague her son, but not with the severity of the previous year.

Then, three months later, more construction work took place in the school – right next to the wall by the son's bedroom. This seemed to unleash a whole new gang of ghosts, and the son and his mother began to see apparitions in the flat again. A shadow appeared by his mother, while another rushed to the side of him and, somehow, violently hit him. The son began to sense a child, and sometimes seemed to be 'possessed', talking in a high, uncanny voice about obscure things that were entirely irrelevant to the family. The ASSAP brought the medium back, who performed further 'rescues' on this new set of spooks. Once more, relative peace settled on the caretaker's flat, and the family moved out of the second-hand caravan in the school grounds and began to sleep inside the building again.

But, like any bothersome infestation, the problem didn't entirely go away. Strange phenomena continued to occur, but with a markedly less violent nature. A 10cm light was spotted hovering around the caretaker and his wife's bedroom, and the son occasionally lapsed into sleep-talking states where his voice sounded unnaturally unlike his own. In wrapping up the case, the ASSAP focused on making sure that the son felt reassured and prepared for any future spooks. As mentioned earlier, he had no prior knowledge of or interest in the paranormal, and the investigators recommended that he start doing some research. They suggested books to help guard himself against further ghostly attacks, as well as learning techniques to hone and control his psychic abilities. Due to his religion, he was reluctant to get involved in Spiritualist circles, but was encouraged instead to speak to members of The Churches Fellowship for Psychical and Spiritual Studies (CFPSS).

Established in the 1950s, the CFPSS aimed to bring Spiritualistic ideas back within a Christian context. As we saw in the previous chapter, Spiritualism began out of Christian ideas but increasingly distanced itself in favour of scientific terminology. But it was always undeniable that theories of spiritual survival after death had their roots in religion, and the CFPSS was set up to explore and study paranormal phenomena through a Christian lens. While in the early years the CFPSS attracted several thousand members, numbers are dwindling in line with declining rates of religion in the UK. Nevertheless, the organisation continues to offer advice and support for Christians in relation to hauntings, paranormal phenomena, and notions of life of death with particular reference to baby loss and grief. There are a number of freely available publications on their website, which are written as practical and relatively rational guides; their pamphlet about dealing with ghosts, for example, encourages percipients to check for natural causes such as faulty wiring or dodgy radiators.[8] The CFPSS' advice for dealing with lingering spectres is to speak casually with them to help them move on: 'It is almost as though we need to say something like, "Sorry, Dad, but you can't hang around. You don't belong here any longer. You've got bigger things to worry about now."' If the caretaker's son did seek advice from the CFPSS, it's likely he would have been encouraged to use his religion to face the ghosts head-on and gently persuade them to leave.

While the origins of nineteenth-century psychical research societies were reserved for the educated middle class, they soon opened up in the twentieth century to be more accessible to people from all walks of life with an interest in peculiar phenomena. These days, anyone can go ghost hunting, for better or worse – as we'll see in Chapter 10. Paranormal investigations have become a bit of a runaway train, with methods becoming increasingly influenced by social media and YouTube rather than organised best practice.

Steve Parsons, an experienced investigator associated with the Society for Psychical Research and the ASSAP, is perhaps the most prominent figure in UK ghost hunting when it comes to attempting to streamline paranormal investigations. As a member of the SPR, he wrote the society's *Guidance Notes for Investigators of Spontaneous Cases*, a 70-page bible that covers everything from ethics to equipment. The purpose of investigations, he writes, is 'to bring some measure of control and measurement to these cases by using established scientific techniques and methods from any relevant scientific discipline.'[9] The guide explains how to separate fraud and natural causes from potentially genuine paranormal phenomena through assessing witness testimony in combination with subjective and objective information. One of the key points of the book is the use of scientific methodology – that is, understanding the phenomena through a hypothesis, and then testing that hypothesis. If a strange noise is heard throughout the house at random times during the day, a reasonable hypothesis could be to suggest that it's caused by movement in the pipes when the boiler suddenly shudders into life. The investigator should then watch for the boiler to become active, and see if it coincides with the noise. Parsons emphasises that the 'scientific' nature of investigations doesn't necessarily mean having lots of complicated equipment; it's more about the steps taken in a scientific investigation to ensure a controlled environment, the testing of theories, and efficient and thorough note-taking. Indeed, while Parsons isn't against equipment, he suggests a fairly sparse approach. Really, a notepad, pen and wristwatch are all an investigator needs, but things like two-way radios, an audio and video recorder and thermometer are also useful. As Parsons writes, 'With regard to the selection of items of equipment for use during an investigation, the investigator must take notice of the fact that currently there is no device which has ever been demonstrated to have the capability to measure or record paranormal phenomena.' An electromagnetic field meter (EMF meter), for example, is a popular tool used by independent paranormal investigation groups who use the coloured lights on the ironically coffin-shaped device to try to communicate with ghosts – they will ask the 'spirit' to make the meter's green light flash to answer questions or indicate their presence – but as Parson notes in his SPR guide, this is not proof of communication with the dead, it's simply showing that there are EMF spikes that coincidentally come after a question is posed to the alleged ghost.

The guidance notes make it clear that every investigation needs to be followed by a clear and concise report, with some attempt at a conclusion (even if the conclusion is that more work needs to be done). What Parsons stresses, particularly in relation to the ethical considerations of a paranormal investigation, is that the investigator's role is simply to gather data and assess whether the phenomena can be explained as having a natural cause or not. Especially where there is a member of the public involved as an initial witness, the investigator must not act as a counsellor or therapist, no matter how much the percipient tries to instigate that dynamic, nor should the investigator provide personal opinions on what the phenomena could be. Parsons also states repeatedly that investigators should not approach the media, although, particularly when we look at poltergeist cases in Chapter 8, this is easier said than done.

While the SPR has Parsons' book of guidance notes, they do not currently offer any formal training for their members. This is where the ASSAP comes in; they are the only society out of the Big 3 that prioritises training for would-be paranormal investigators. Anyone can join the ASSAP, but to assist in their cases members must undergo at least the first module of their two-part training scheme, delivered by Steve Parsons himself. Naturally, I signed up.

On a very rainy and atmospheric day in late April 2024, I drove to The Old Prison in Northleach, in the Cotswolds, to become an ASSAP Accredited Investigator (AAI). As we saw earlier, the group answer calls from all over the country for help with alleged hauntings and other phenomena. From warehouses to studio flats, the ASSAP is brought in by residents and property owners when they experience things they simply cannot explain. On a voluntary basis, and for as long as is deemed necessary, ASSAP investigators conduct a thorough examination of the property. Their mission is simply to try to gather evidence as to whether the phenomena can be replicated, and whether there is a natural explanation for the strange occurrences. Steve Parsons is the ASSAP's official training officer, and as he explained to me and my fellow trainees, often people call the ASSAP simply to have their experiences taken seriously. Most of the time, simply having a team investigate and try to recreate phenomena is enough for them; they don't mind living with the phenomena as long as they know that there are people out there who don't think they're going mad.

As an AAI (which I now am, congratulations to me), I form part of a nationwide register. If anyone in Ceredigion experiences a spook and seeks the ASSAP's help, I would likely be called upon to conduct the investigation – unless the case involves a UFO, of which I have a bizarre phobia, and then I would politely but firmly decline. For most of the day I spent training, I learnt about how to ghost hunt properly, using ethical methods of fact-finding. I sat down and opened my notebook, purely interested in gathering notes for this book, but I soon found myself drawn into the world of paranormal investigations. I was fascinated by the psychology of it all, by the way in which an entire investigation could change simply by naming the phenomena a 'ghost'. Indeed, Steve related a somewhat botched case by example of what *not* to do. A homeowner had called the ASSAP to ask them to come and investigate strange phenomena. They were fairly nonplussed, more intrigued than anything, but as soon as one investigator said to the other, 'This reminds me of that poltergeist case, do you remember?' the homeowner became frantic with worry and nearly moved out of their home. It struck me that the focus of the ASSAP's current investigations is on language – how to phrase phenomena without leading the percipient to jump to conclusions about possible paranormal causes, but equally how to use language in a way that suggests belief in what the percipient says they experienced. If they say they saw a ghost, then call it a ghost. But if they haven't come to a conclusion on the phenomena, don't scare them by saying, 'Oh boy, that's definitely a poltergeist – good luck!'

The training sessions also seemed to draw attention to the current state of paranormal investigations, and the pervading influence of the British reality ghost-hunting show *Most Haunted*. We'll look in detail at this aspect of ghost-hunting history in Chapter 10, but it struck me that amateur paranormal investigators are the absolute bane of the ASSAP's existence, and part of the training was to try to organise and control this otherwise very disorganised activity. Several of my fellow trainees were just such amateur investigators, and I often got the sense that they bristled at some of the advice we were being given. This was even more apparent when we went on our practice ghost hunt in the evening.

The Old Prison was built in 1790, with a design that implemented the new prison reform guidelines that Sir George Onesiphorus, the high sheriff of Gloucestershire, stoutly believed in. Cells were built in a hexagonal shape so that prisoners could be held securely without being fettered, although most of the building was demolished in the 1930s leaving just a few rooms, the courtroom and a handful of cells in the main part of the building. We were split into three groups, and the Old Prison was divided into three areas. For an hour, each group spent time investigating one of the three locations before changing to a different area.

My group went down to the cellar first, from seven until eight in the evening. It was a cold, cramped place where sound seemed to be magnified. The sound of a motorbike speeding down the road outside roared through the tight space. We stood in various places, where we could all see each other and cover every inch of the cellar. The ASSAP's instructions for investigating were to stand quietly and write regular notes about any feelings we had or observations we made. One member of my group, an amateur paranormal investigator whom I'll call Astrid, told us all how strange this was for her – she was used to having various bits of equipment, like an EMF meter, that she used to communicate with spirits. Unlike her usual methods, we didn't call out or ask the ghosts to come towards us; we stayed quiet, observing.

Mostly, I observed the large dead spider that dangled precariously above my head where I had chosen to stand.

At the same time as I spotted this, Astrid said she suddenly felt chills – but not from the cold. It was as though something walked through her, she said. Five minutes later, I suddenly had an urge to brush something off my head. I looked up in horror, but the dead spider was still dangling there. It felt as though someone had cracked an egg on my scalp, but equally I'm the sort of person who has visceral, bodily reactions if you so much as mention a spider, so I'll chalk that up to my paranoia.

There were some interesting outcomes, however. There were six remaining cells in the prison, over two levels, and my team split up so we each took a cell for a few minutes, and then rotated. What I found fascinating was how each cell had its own *feel*. One was colder than another; one was darker and damper than another. In one cell, I felt a little off-balance, in another I felt strangely lonely. In another cell, I wrote a particularly astute observation in my investigation notes: 'It's giving Year 8 History fieldtrip vibes.' Upstairs, above the courtroom, there was a strange rabbit warren of rooms, small passages and staircases that led nowhere. I sat with Astrid's fellow paranormal investigation group member, Dora, in a very dark room. I could only just make out Dora's outline as she sat on a chair a few metres away from me.

I was incredibly cold, colder than I had been in the cellar or any other part of the prison during the night, and shivered for the first part of this investigation, burying the lower half of my face in my giant scarf. Then, almost as though I'd been plunged into a nice bath, the cold lifted and my muscles relaxed. I felt positively warm. I turned on my torch to write this down in my notes, and as soon as my notepad became illuminated, Dora told me she suddenly felt freezing – as though whatever had been making me feel cold had passed on to her. I had said nothing about what I was experiencing, and hadn't even begun to write down my note, which, even if I had written it down, was too far away from Dora for her to see my spiky handwriting at all. It could, of course, have been a purely subjective coincidence, and since we didn't have thermometers to hand we couldn't back up this experience with data. Nevertheless, it was a bit weird.

After this final vigil, the three teams reassembled in the main function room of the Old Prison to go over our findings. What I found interesting was that there wasn't a person in the room who hadn't experienced anything at all. As we discussed each location, the variety of sensations and phenomena was quite staggering, from headaches to tingling limbs to mysterious breezes. Everything was, however, subjective. Even though Dora and I felt that we had 'shared' an experience, it wasn't a properly corroborated report; we didn't both see a light or an apparition, but simply felt simultaneous and opposing temperature changes, which was more likely coincidental than proof of anything spooky.

The next day, anyone I'd told about the training asked me if I'd seen a ghost. I had to disappoint them, but then that's surely the whole point of ghost hunting. If everyone encountered an apparition the first time they went on an investigation, we'd have solved the question of the existence of ghosts a long time ago. This is why we keep searching, because we find ourselves experiencing something a little strange, a little tantalising – like the way I went warm as Dora went cold – that makes us wonder if, next time, we'll encounter something even stranger. There always has to be a next time.

CHAPTER THREE

Haunted Houses

In the early days of Spiritualism, the focus of investigations remained largely on séances and materialisation mediums, summoning spirits to rooms often far removed from their homes. As ghost-hunting committees became established, however, they sought to examine a wide array of phenomena. While the debate surrounding séances and mediums still raged around them in the late nineteenth century, the SPR in particular wanted to look more broadly at psychical anomalies and alleged cases of ghosts, and soon expanded their inquiries in a way that returned ghosts to a much more traditional setting. Immediately upon forming the society, a special committee was assigned the task of investigating location-based hauntings, focusing on ghosts in their natural habitats of houses, castle ruins and graveyards, and their work permanently influenced the way ghost hunts took place in buildings.

In the introduction to this book, we looked at the origins of ghosts and ghost stories – particularly the revenant spirit wandering around the site of their death or where they resided in life. Even amid the buzzing popularity of Spiritualism and séances, the traditional ghost stories pervaded with regularity. It was these to which psychical researchers returned, particularly as the twentieth century dawned, with this shift of interest really signalling the birth of true ghost hunting. Members of psychical research societies and independent investigators took to the sites of famous and recent hauntings, mimicking the tactics and procedures of the relatively recently organised national policing structure in Britain. They were given tip-offs, searched for clues, staked out buildings at night, interviewed witnesses and kept fastidious notes on anything that could be considered ghostly phenomena.

One of the most substantial cases from this era was published in the *Proceedings of the Society for Psychical Research* in 1892. Titled 'Record of a Haunted House' by Miss R. C. Morton, it is a report on experiments conducted in her family home under the remit and guidance of the Society's senior members.[1] It details a chronic haunting, the phenomena of which were mostly represented in the form of a materialised spirit: a stereotypical female ghost in mourning costume, who was in a constant state of weeping and permanently covered her face with a handkerchief.

The house was only a little over 20 years old when Morton began her investigation, an interesting departure from the belief that ghosts inhabited ancient historical sites. Nevertheless, for such a young property the house had already witnessed a great deal of history befitting any castle or old manor house – including tragedy, alcoholism, depression, hidden jewels and death. The story goes that Mr S., a

British-Indian man who was the house's first resident, took to heavy drinking after the death of his wife. Eventually he remarried, but the melancholic atmosphere and shadow of the first Mrs S. caused his new wife to turn to alcohol, too. They argued frequently over their children and the first wife's collection of jewellery. When Mr S. died, his second wife moved away until her own death in 1878, and the house was sold to a Mr L. Unfortunately, however, Mr L. also kicked the bucket after six months – in the very same room Mr S. died. Despite dying away from the house, the second Mrs S. returned in spirit form, forever searching for that elusive jewellery collection.

In the report's prefatory note written by F. W. H. Myers, there is a paraphrased comment from R. C. Morton's father, Captain Morton. The legacy of living in a 'haunted house' seems to have troubled him; he was thinking, perhaps, of the infamous Cock Lane ghost of 1762, which brought crowds swarming from across London and causing disorder around an alleged haunted building. He was worried that their house would depreciate in value, either because of the haunting or because prospective buyers, too, would worry about attracting the 'wrong sort' to their doorstep. It seems it was with reluctance, and only out of kindness to his daughter, that he allowed the SPR to publish the report. It was on the condition, however, that they reported the phenomena to have ceased in the house.

The house sounds rather grand, with a carriage-sweep and orchard, and by the time the Mortons moved in at the end of April 1882, it had been spruced up and was generally in a good, pest-free condition (apart from the ghost, obviously).

The first encounters with the spectre of the second Mrs S. come, like all good ghost stories, via neighbourhood hearsay. A gardener was said to have seen a lady wandering around; another woman, a town local, insisted she had also seen an apparition shortly after the death of Mrs S.

By June, two months after they arrived, Miss Morton saw the apparition for the first time. She writes of her initial encounter:

> I had gone up to my room, but was not yet in bed, when I heard someone at the door, and went to it, thinking it might be my mother. On opening the door, I saw no one; but on going a few steps along the passage, I saw the figure of a tall lady, dressed in black, standing at the head of the stairs. After a few moments she descended the stairs, and I followed for a short distance, feeling curious what it could be. I had only a small piece of candle, and it suddenly burnt itself out; and being unable to see more, I went back to my room.

One thing's for certain: Miss Morton was braver than me. Then again, she was from a large middle-class family that presumably often received guests and visitors. Perhaps, in her situation, it wasn't that uncommon to see a strange woman walking around the house. Still, we have to admire her for attempting to follow the figure, only retreating when the stub of her candle guttered out.

Despite mentioning in her initial encounter that she was viewing the figure in dim candlelight, she describes the woman in fairly extensive detail. The figure was tall, dressed in black mourning garb – Miss Morton heard the rough woollen rustle of the fabric as the woman moved – and hid her face in a handkerchief.

Miss Morton would repeatedly see the apparition over the next two years, in various places and at various times of day. While her first encounter took place in the dark, she just as often saw the weeping woman in the daytime. She claimed to have seen the figure about half a dozen times, at increasingly short intervals of time, as though the woman's power and presence over the household was growing.

During these two years, only three other people claimed to have seen the apparition – and even then, only once. Miss

Morton's sister saw the figure in late summer, not long after the initial encounter described above, and mistook her for a nun who was visiting the house. A year later, a housemaid raised the alarm around bedtime that there was an intruder in the house, her description of the figure matching that of the lady in mourning dress. Finally, a few months later, in December 1883, Miss Morton's six-year-old brother and his friend were playing outside and saw the spectre of Mrs S. in the drawing-room window. I'm not sure I would have included the testimony of a child, though; when I was of a similar age, playing in the garden at a friend's house, we thought we saw a ghost peering out at us through her bedroom window. Terrified, we gathered up all our courage and mutual love of *Scooby-Doo* to go back inside and confront the spectre. It turned out to be my friend's white dressing gown hanging on the door.

Since the ghost appeared most frequently to Miss Morton, she seized her chance to conduct investigative work. She kept journal entries and wrote letters to her friend documenting the phenomena. By doing this, she noticed a pattern; Mrs S. seemed to appear at the top of the stairs, then walked down to the drawing room, where she looked out of the window to the right for a while before disappearing down the passage towards the garden.

By 1884, Miss Morton became rather used to seeing the apparition. If she ever felt any fear (which she never admits to), by this point her feeling towards the ghost was pure curiosity. She attempted to communicate with the ghost, not through table-tipping or a séance but by simply speaking to the apparition and asking if the woman needed help. On one occasion, the apparition turned to Miss Morton and gasped, but did not respond. This alone is fascinating. Séances were still very much in vogue at this time; why didn't Miss Morton attempt to communicate with Mrs S. through such a popular method? Surely her siblings would have been on board, even if her father, Captain Morton, was sceptical. Moreover, the ghost acted like a sleepwalker, repeating the same pattern of activity over and over again, walking from the top of the stairs to the garden. Yet, here, she suddenly became lucid and sentient, a different kind of spectre altogether – one who is aware of her environment and the living people around her – which would have made her ripe for séance communication.

There's something about Miss Morton's experiences that takes us back to the Toronto Society for Psychical Research's infamous Philip Experiment. In reflecting on the phenomena created by the group's projection of their fictional ghost, Iris Owen and Margaret Sparrow wondered how their findings related to paranormal encounters throughout history. They examined the role of *expectation* as a cause of phenomena, and described a situation that seems particularly pertinent to Miss Morton's case: 'The family ghost in the haunted mansion appears always on the same staircase, in the same dress, and with the same expression, points in the same direction, because the family members all know that this is just as it should be.'[2] Interestingly, Captain Morton and his wife never encountered the ghost; their scepticism blotting out any expectation. There was even an occasion, in July 1884, where Miss Morton was sitting in the drawing room in the evening with her father and sisters, and saw the ghost of Mrs S. make her usual journey past them all to stand behind the couch where Miss Morton had positioned herself to read her book. 'I was astonished', she wrote, 'that no one else in the room saw her, as she was so very distinct to me.' She claimed that the ghost stayed behind her for *half an hour*, an exceptionally long time, and still neither her sisters nor her father could see what Miss Morton could see. Perhaps, as with the group involved in the Philip Experiment, the hearsay around Mrs S. had created a certain level of expectation in Miss Morton that no one else in her family shared to the same degree. She clearly enjoyed writing, as shown in the journal she kept and the lengthy, detailed letters she sent to her friends, and was thus a creative individual. Was she able to see the ghost through some innate psychic ability, or was the haunting all a product of her vivid imagination?

As the twentieth century approached, haunted houses became even more of a focus for ghost-hunting societies, scientists and Spiritualists. This was, after all, a period in which a fascination for Egyptian culture and archaeology swept through the British middle classes, with elaborate expeditions uncovering the tomb of Egyptian queen Nefertari in 1904 and, famously, Tutankhamun in 1922. Haunted houses offered the same thrill of adventure, discovery and, of course, occult encounters, but without the extravagant expense of a lengthy overseas voyage.

They also allowed anyone, from any discipline, to comment and participate in investigations, whereas Egyptian archaeology was reserved for those who specialised in the subject and its techniques. In 1924, for example, the influential French astronomer Camille Flammarion published *Haunted Houses*, a book about his own thoughts and investigations into ghost-infested buildings. Like William Crookes and Sir Oliver Lodge, Flammarion was a key figure in the debate between science and Spiritualism because of his prominence in his own field of study. And, like Crookes and Lodge, despite his legacy and discoveries, he was also a staunch believer in the paranormal and was particularly interested in haunted houses. After all, astronomy as a science was based on the discovery of things that had always existed but simply needed the right equipment to be seen – so too did ghost hunting simply require theories, patience and the right tools to find what had previously been elusive. For over 50 years, Flammarion had been conducting investigations into haunted houses and the testimony of others, which he felt qualified him to 'affirm here, somewhat crudely perhaps, but clearly, that the people who jeer at stories of haunted houses and deny their reality suffer from a special form of myopia, so that their horizon does not extend much beyond the tip of their nose.'[3] He proposed a scientific theory of haunted houses that wouldn't be out of place in some of his astronomical works; he argued that any solid object can act as a recording device for what he called 'effluvia, residual forces, and vital fluids', which, when touched by the right sort of person in the right sort of circumstance, causes the effluvia to be released in the form of a ghost or paranormal phenomena. 'The walls and furniture' of a haunted house in particular, 'may preserve the imprint of events with which they were associated.' The idea was later popularised as 'Stone Tape Theory', from the 1972 BBC Christmas horror film and ghost story *The Stone Tape* by Nigel Kneale – which proposes that ghosts and related activity are actually impressions recorded in stones in the walls, and are randomly spat back out again as though playing a vinyl record on repeat – and remains very much in the vocabulary of modern investigators.

In the history of ghost hunting, however, no haunted house investigation is perhaps as notorious as that of Harry Price's residency at Borley Rectory, Essex. Harry Price (1881–1948) is perhaps best described as a professional hobbyist; he was an author, but had a vested interest in antiques, collecting, classical history and archaeology. He also loved ghosts. He joined the Society for Psychical Research as a young man in 1920, but after a few years became something of a rogue figure in these circles, going on independent investigations and taking the glory for himself while always keeping an eye on his rivals.

Borley Rectory was the perfect opportunity for Price to conduct a long-running series of tests and experiments that, he thought, would see him become the world's top ghost hunter. In 1929, the site became known to the public through a series of *Daily Mirror* articles discussing its alleged haunting. As Price told it, the editor of the newspaper, Mr V. C. Wall, called him up to ask for his professional assistance in investigating the 'remarkable incidents' that were taking place at the rectory.[4] Price clearly saw an opportunity for independent fame and success, and set off the next day with his secretary Lucie Kaye and a bag full of ghost-hunting equipment.

Arriving at Borley, they were received by the current occupiers, Reverend G. E. Smith and his wife, who were only too eager to tell Price and Kaye about the lengthy catalogue of phenomena they had experienced during their short time there. The Smiths had only resided at Borley since late 1928, less than a year, and told Price that the rectory had been refused by 12 other clergymen before Smith agreed to move in – but only because he hadn't been told of the rectory's rich and troubled history.

The first ghost stories of Borley and the land surrounding it were around 700 years old, and in time more and more spirits were sighted and became legends attached to the site. Before Borley Rectory in its modern form was built by the Reverend Henry Bull in 1863, there existed Borley Monastery. One story goes that a monk from the monastery ran off with a nun from the nearby Bures Nunnery. Their elopement was short-lived and they were soon caught, and the tale ends with a macabre fate for them both: the monk was either hanged or beheaded, and the nun was returned to Bures and bricked up alive. There is a second story, however, in which another, unrelated nun met a grisly end. This time, a young French nun was forced to leave her nunnery in Le Havre and marry a man from a prestigious Borley family. She was strangled by him in 1667 and buried beneath the land that later became Borley Rectory.

The gruesome death of a nun, in one form or another, seemed to be the principle revenant at Borley, although there is no proof of a nunnery at Bures, nor of a murdered French nun. While the story of the illicit romance remained legend in the surrounding community, the building of Borley Rectory by Rev. Henry Bull rekindled old ideas of restless, vengeful, tortured spirits, and it was during the Bulls' occupancy that the modern tales of haunting pervaded. The family would frequently hear strange sounds and disembodied footsteps and, most chillingly of all, would look up from the dinner table to find a spectral nun peering at them through the window.

Price was intrigued; already there were strange discrepancies and mysteries surrounding the story. In one instance, he heard that Bull had bricked up the dining room window so they couldn't see the nun staring at them, but was later informed he'd actually done it because trespassers and ramblers were coming close to the rectory and peering in on the family. But Price was skilled at believing whatever would bring him the most publicity, and took what he had first heard to be the truth: that the bricked-up window kept out the spectral nun. He was also aware of the dubious nature of the original legends, and how neither story contained any proof, but he chose to make it his mission to find that proof for himself.

Price began his investigation proper on the afternoon of 12 June 1929 by drawing up a detailed floorplan of the building, complete with measurements of furniture and fittings. I mentioned earlier that haunted-house investigations mimicked Egyptian archaeology; Price's work at Borley also has a smattering of wartime preparation and strategy about it, too. Together with his secretary, they navigated through tight, dusty cupboards and cubby holes, checked the walls for cavities and draughts, examined the rafters and attic, and sealed all the windows and doors of the upper storey to prevent anyone getting in and contaminating the site. They then checked whether they could easily move particular items of furniture themselves, and looked at the rectory's bell system. This, Price found, had mostly been disconnected to stop 'the incessant ringing' by ghostly hands. Down in the cellar of Borley, they encountered a scene straight out of a children's horror story: the place was crawling with frogs, toads and newts. While the presence of animals could cause some of the noises and movement at Borley, unless these critters stood on each other's backs and donned a habit, they couldn't really be used to explain the apparition of the nun.

And it wasn't long before Price encountered her. On his very first overnight group vigil at Borley, accompanied by a friend named Mr Wall, he took up position by the summerhouse to watch over the infamous 'Nun's Walk' portion of the garden, where the ghost was said to roam most frequently. Suddenly, Mr Wall cried out, 'There she is!' and pointed towards the trees. Tall and gloomy in the summer twilight, the trees were cloaked in shadow. But one shadow appeared to be darker than the rest, and it seemed to move. Price saw it 'gliding towards the end of the garden and the little stream,' but then teases us with uncertainty, wondering if it was the semi-darkness that played tricks on his eyes or whether he was simply latching onto anything in order to corroborate Wall's experience. I find this happens quite a lot on ghost hunts; one person will see or hear something, and the rest of the group will say they did too, to provide back-up. But when they compare notes, the phenomena don't always match up. For example, someone will say they heard a noise and another will describe it like a baby's cry, only for the first person to say it was more like knocking on wood.

No sooner had they seen the mysterious figure than something even more troubling occurred. Walking back to the rectory in the dark, they heard an explosive crash and dodged out of the way of a shower of glass. A window had been smashed above them. If you're picturing this scene in your head, you'll have realised that for the glass to fall on Price and Wall *outside*, the window would have to have been broken from *within* the room itself. Sure enough, they found a lump of brick on the veranda which they took to be the missile. Given how meticulously Price had searched the building and grounds all afternoon, he likely would have come across the brick before. Inside the now-draughty room, things became even stranger: nothing was out of place and the seals around the doors and windows were intact, suggesting that the rest of the group who'd remained inside the rectory hadn't lobbed the brick out of the window themselves. While discussing the broken window with those who were holding vigil inside, a red glass candlestick came hurtling down the stairs and, like the window, shattered into jagged pieces. Next, a mothball bounced down each step and hit Mr Wall's hand. But everyone was accounted for; no one was upstairs. Yet objects kept pelting them: 'some common seashore pebbles; then a piece of slate, then some more pebbles'. The service bells began to ring, and Price described how he could even see the wires moving, despite the circuits being disconnected. Finally, in the corridor, both the library and drawing-room keys simultaneously slipped out of their keyholes and onto the floor.

It was an eventful evening. Even discounting his maybe-maybe-not glimpse of the infamous nun, Price knew at this point he had struck ghostly gold.

By July, the Smiths had had enough of Borley. Claiming it was too damp and was affecting their health, and not because they were scared of ghosts, they left for a more hygienic residence in Long Melford. Between 1930 and 1937, the rectory came on and off the market, changing hands multiple times, as restless as the ghosts it housed.

Each time it was made available, it was cheaper. And the more it came down in value, the more people encouraged Price to seize the opportunity to purchase it. One correspondent even suggested that Price should convert the rectory into a nursing home for retired Spiritualist mediums. Price declined, not because he thought it was an absolutely ludicrous idea but because it would cost too much to refurbish it to a condition fit for habitation. If he was to buy it, he would not change it anyway; it would be kept as a haunted house, expressly for investigations.

Finally, he decided to take the plunge. By 1937, Borley remained empty but was now owned by the Reverend A. C. Henning, who lived instead at the much more pleasant Liston Rectory. Price wrote to Henning and asked him to lease Borley for as long as necessary to complete his investigation to his satisfaction. They agreed for Price to rent the building for £30 a year, bills included, which would equate to about £200 a month in today's money. I've definitely rented mouldier places for three times that sum, so he got an excellent deal.

Price's ghost hunt at Borley began in earnest once the building was his to use as he pleased. He decided, first and foremost, that he didn't want to be involved in the investigations. I find this quite perplexing, as I would've been snooping around the house all the time if it were me, but I suppose I understand where he was coming from – he was trying to avoid becoming too emotionally invested in the building. He wrote in *The Most Haunted House in England* that he wanted '*independent evidence* from intelligent, competent, and cultured strangers who were *not* spiritualists, and, if they knew nothing about psychical research, so much the better' (original emphasis). He began to recruit his cultured strangers, and on 25 May 1937, Price's advertisement in the *Times* was printed:

> HAUNTED HOUSE. Responsible persons of leisure and intelligence, intrepid, critical, and unbiased, are invited to join rota of observers in a year's night and day investigation of alleged haunted house in Home Counties. Printed instructions supplied. Scientific training or ability to operate simple instruments an advantage. House is situated in lonely hamlet, so own car is essential.

He wasn't asking for much, then.

One thing I've noticed in terms of the history of haunted house investigations in particular is the gradual shift in accessibility. In the early and mid-twentieth century, Price and his ghost-hunting colleagues were middle- to upper-class citizens. They were people who could afford the fees of psychical research society membership and specialist equipment such as cameras and microphones, and who were, as Price so delicately puts it, 'persons of leisure', who were financially comfortable enough to spend time chasing phantoms in a dusty old house rather than be at work. Now, paranormal societies are dominated by working-class people who go on ghost hunts in their spare time; it is no longer a hobby of bored and wealthy people, but an active way for people from all backgrounds to be part of a community experiencing allegedly haunted locations.

Price received around 200 applications in all, mostly from said bored and wealthy people. There's a sense that not many people took his advertisement seriously, which is perhaps a sign of the decade's commercialisation of Ouija boards and sensational pulp horror books. Some, Price remarked, even had the temerity to ask for payment. Others told him he was dabbling in dark forces and offered to exorcise the house instead of investigate it. Nevertheless, after much sifting through these letters, Price hand-picked a selection of people whom he thought fitted the bill.

Next was the matter of training. Each person was interviewed by Price, and a declaration form was signed that agreed to a number of conditions including participants covering their own expenses, not working with any newspaper or journalist, not revealing the location of the site (although, as we've seen, Borley had been in the papers more than once – someone may well have been able to make an educated guess), not photographing or sketching any part of the house, not allowing pals in to join the vigil who hadn't signed their own declaration form, and, quite crucially, not writing or lecturing about anything they had experienced during their time at Borley. Price wasn't going to let anyone steal his publishing deal.

For the purposes of the investigation, and to ensure that his unpaid research assistants were streamlined in their methods of inquiry, Price drew up a guide which he proudly called 'The Blue Book, Instructions for Observers, Private and Confidential'. Like the declaration form, this also listed a number of strict and, at times, meticulously prohibitive rules to follow. The Observers were to carry the following equipment: 'note-block, pencils, good watch with seconds hand, candle and matches, pocket electric torch, brandy flask, sandwiches, etc. If he possesses a camera, this can be used. Rubber- or felt-soled shoes should be worn'. Lights were to be extinguished, doors fastened and meals to be taken at scheduled times. Price had kindly left a single camp bed at Borley for 'sufficient rest' to be taken (by which I think he preferred none were taken at all). In total, 48 Observers passed Price's interview and took shifts to investigate the haunting of Borley Rectory.

During the time of the Observers, one particular manifestation stood out: strange markings, seemingly made by a pencil, appeared spontaneously on the walls. This had been going on since the Foyster family had taken residence in the rectory earlier in the decade, but was now continuing in earnest, sometimes right before the Observers' eyes. In the reports sent to Price and his new secretary, Mollie Goldney, some Observers copied down the symbols and words. One page is full of loose 'B' shapes that were traced as they appeared on the wall, and other reports speak of the messy, frantic messages regarding a girl named 'Marianne' which had first been spotted by the Foysters. Two of the observers, Mr Glanville and his brother-in-law, were taking photographs of the space in which Marianne's messages had appeared. An hour after taking their first photograph, they noticed more messages had been added and so took another. Oddly, however, on developing both they found that only one of the photographs captured any of the messages at all. Price got a little too excited by this, and said, 'If this is not *proof* of paranormal activity, I do not know the meaning of the word.' But since Price wasn't there with Glanville, how could he know for certain that the two men hadn't taken a photograph of bare wall and then added their own messages before photographing it again? The more certain Price was of the spurious evidence produced by his Observers, the more he raised the hackles of the rest of the psychical research community, who had very strong opinions on the whole affair. And, indeed, betrayal would come from those within his inner circle of ghost-hunting colleagues.

But, for now, the investigations continued in earnest. The Observers recorded loud bangs and crashes without an obvious cause, and one team decided to bring in a cat to see if it reacted strangely to any particular spot in the rectory (it didn't). Despite his initial rule not to investigate Borley himself at this time, Price continued to pay visits to his own haunted house. Intriguingly, on the final night of his tenancy in 1938, as he kept watch with his friend Mr Motion, a 22-carat gold wedding ring that belonged to no one materialised in the air and dropped on the floor of the 'Blue Room'. Mr Foyster, seven years prior, had also witnessed the same phenomenon, but the ring had disappeared the next morning.

Two months before Price handed the keys to Borley back to Mr Henning, Mr Glanville's daughter, Helen, had been dabbling with planchette séances and decided to give it a go while her brother was also visiting home. The planchette was a commercial spirit-communication tool that was widely available in bookshops from 1868; it was a teardrop-shaped piece of wood on casters, often with a hole in which a pencil could be inserted.[5] Séance-sitters would lightly touch the wood, and a spirit would allegedly push the planchette either over a Ouija board or, with the pencil attached, on a piece of paper to write messages. In Helen's séance, the planchette immediately began to move, the pencil scribbling furiously. When it finally stopped, Helen and Roger lifted up the planchette to reveal the following ominous message that Price included in his report:

> *Sunex Amures and one of his men* [indistinct] *MEAN TO BURN THE RECTORY to-night at 9 o'clock, end of the haunting go to the rectory and you will be able to see us enter*

> *into our own and under the ruins you will find bone of murdered* [indistinct] *wardens* [not clear] *under the ruins mean you to have proof of haunting of the rectory at Borley* [indistinct] *the understanding of which* [indistinct] *tells the story of murder which happened there.*

The rectory did not, in fact, burn down that night. But it did burn down. It wasn't until nearly a year later, on 27 February 1939, that a lamp was a knocked over in the hall and started a blaze that entirely devastated the building.

Does a haunted house stop being haunted when it's burnt down? In Borley's case, apparently not. A month after the terrible fire, a group of locals were taking a moonlit walk around the charred remains of the rectory. In what was left of an upstairs window, the group distinctly saw a woman dressed in blue peering back at them. She remained where she was for a few, long seconds, and then glided away, vanishing into the wall. The interior of the rectory was completely inaccessible; even if they saw a flesh-and-blood woman, it would have been impossible for her to have got past the safety barriers that had been put up, let alone navigate the crumbling insides. Even stranger, the floor beneath the window had disintegrated, so the woman couldn't have been standing on anything.

What Price thought was a closed case, with a successful monograph on the investigations, still had more to reveal. Vigils began again, and, bizarrely, the most rigorous group to visit the rectory included several members of the Polish Army Medical Corps. In June 1943, Lieutenant G. B. Nawrocki sent Price a thorough and detailed report, to the minute, of what he and three of his colleagues had experienced over two nights. The haunting was very much still active. The men encountered stones being thrown at them, doors opening and closing of their own accord, and at 11.05 p.m. on the first night of their investigation, Nawrocki saw 'a black shadow moving slowly between the trees on the Nun's Walk'.[6]

But time was running out for Borley. Its current owner, Captain Gregson, was due to sell off the property, after which, in all likelihood, the remains would be razed to the ground. Price grasped his last chance to hunt Borley's ghosts, this time by excavating its cellars. Throughout his investigations at the rectory, what lay *beneath* the site was alluded to again and again in local legend and in the occasional séance. Not only had the land served other purposes before Borley, but there was apparently a plague burial pit and, according to the spirit Sunex Amures, the location of the bones of a murder victim. Price's team (his heart complaint stopped him from doing any digging himself) did, in fact, find bones, but whether they were human or pig is unclear.

Borley had one final piece of evidence to reveal to Price. A man named Mr Scherman was among the crew excavating the cellars, and had brought along a nifty little American camera in order to take documentary photographs of the ruins and the ongoing excavation. Standing next to Scherman, Price watched as he took a photograph with a wide-angled lens of the entire building. Suddenly, in an echo of one of the first phenomena Price encountered at Borley so many years before, a lump of brick 'shot up about four feet into the air'. It was only when Scherman developed the photograph that he realised he had actually captured the brick in mid-flight.

It became a somewhat infamous photograph, and created appropriate publicity for *The End of Borley Rectory*, Price's follow-up book to *The Most Haunted House in England*. It wasn't, in fact, the end at all. Where Price had declared himself officially finished with what he considered to be the pinnacle of his ghost-hunting career, his rivals were only just beginning. It took them 10 years, in which time Price was dead and buried, but Eric Dingwall, Trevor Hall and, devastatingly, Price's own secretary Mollie Goldney finally published their detailed exposé of the Borley investigation. They innocently called it *The Haunting of Borley Rectory*, but the contents are largely a critical interpretation of Price's reported evidence.

Dingwall and Price had sustained a long-running feud, but more complex is the involvement of Goldney. It was something that confused me when I read the Borley books, but it was only through reading Goldney's letters in Senate House Library in London that I understood a little more of her character. There is a stark difference in the way she writes to both Dingwall and Price. Her letters to Dingwall are informal; they are like brother and sister, teasing each other, confiding in each other, sharing gossip (mostly about Price). In direct contrast, there's something cold and long-suffering in the way she wrote to Price, not least because Price's letters to her often asked her to do ridiculous administrative tasks like single-handedly organising a club dinner with incorrect or very little information. She may have been employed by Price, but her loyalties lay with Dingwall and his idea of proper psychical research.

Subsequently, Goldney was personally invested in taking Price, and his Borley books, down. In a letter from Goldney dated 14 July 1937, she told Dingwall that she was off to spend the night in Price's 'haunted house' – she didn't reveal the name, but her description of the phenomena make it obvious that it was Borley.[7] Dingwall, in his reply, jokingly warned her not to 'let the spook camel be sick on you.'[8] What a bonkers turn of phrase. Goldney followed his advice, and found the decaying building to be devoid of ghostly activity despite Price's best efforts to conjure up spirits. While there, she had encountered phenomena that she believed was being caused by Mrs Foyster; Price didn't investigate this probable human explanation when Goldney prompted him, presumably fearful it would cause his entire operation to collapse. She took none of it seriously while it was ongoing, and was even more irked when Price began to publish his writing on Borley. This is inherently clear in a letter she wrote to Dingwall in December 1940 to discuss *The Most Haunted House in England*, in which she gave a rather biting review: 'I have not read it, because it is not worth its cost and I haven't found anyone to borrow it from.'[9]

She did, however, read it eventually – only to pull it to pieces. In fact, together Dingwall, Goldney and Hall disembowelled nearly everything about Price's investigations, from his choice to employ the Observers from the general public to the way in which he was easily taken in by unsubstantiated local legend. In particular, they noticed that apart from the sightings of the 'nun', the phenomena recorded by Price and his Observers were so wildly different and more violent than anything previously described by Borley's former occupants. It was all performance, all melodrama, and despite Price outlining that he didn't want his Observers to be members of the press, he clearly wanted the attention of the British media. Indeed, ironically, he allowed several BBC staff to conduct investigations at Borley. His 'besetting sin', claimed Dingwall, Goldney and Hall, 'lay in his passionate desire for publicity'.[10]

It was this desire for publicity which led the authors to their damning accusation: that Price both faked and deliberately obscured evidence to make the investigation appear much more successful and evidential of genuine paranormal activity. They criticised the Observers, who lacked experience and training, and were concerned by the fact that each Observer started the investigation afresh. There was no continuity; one Observer didn't take up where the previous one left off. They didn't know what each Observer had already done or tested or seen. And because of that, there were giant holes in the reports, as well as numerous glaring discrepancies. The most troubling was that of the 'Marianne' messages, which some Observers noticed and some did not. Other messages appeared for one Observer and then were not present for another. All the while, Price presented their evidence as absolutely trustworthy.

In short, the authors wrote that it was 'possible to regard Price as a brilliant if cynical journalist' – a journalist, not a ghost hunter – 'who used the material gathered either in his laboratory or in the field in such a way that its publicity value was highest. As we have seen, if the material lacked sensational elements it would seem that he was prepared at times to provide these himself.' Worst of all, the famous photograph of the flying brick was only a section of a larger image. In the original, a man can be seen in the far left, hand still raised, the brick having clearly originated from him. But, unfortunately, this new revelation wasn't known by the time *The Haunting of Borley Rectory* was written.

Dingwall, Goldney and Hall's book didn't quite deal the blow to Price's career they had hoped it would. In fact, in some respects it had the opposite effect. Perhaps if they had published it immediately after Price's books were released, it would have been a different matter. However, when *The Haunting of Borley Rectory* was eventually published, Price had already been dead for eight years. The buzz around Borley had long been dead, too. The distance provided by time, and the fact that Price wasn't around to defend himself, meant that when his detractors' book was released in 1956, it was seen as sour grapes. Why wait until now to pick apart the Borley reports? Why do it at all when Price was long dead? Did anyone care about the veracity of the experiments but Dingwall, Goldney and Hall themselves?

Moreover, Harry Price's wife was still alive. The Hungarian journalist and keen member of the British Psychoanalytical Society, Paul Tabori, wrote to Eric Dingwall on 13 January 1956, at which point the publication of *The Haunting of Borley Rectory* was imminent. Advance reviews and articles were already being printed in newspapers, including one in the *Sunday Express*. Things weren't sitting well with Tabori, who put in his letter:

> This has very much upset Mrs Price and caused her considerable distress. I wonder whether, for her sake, it would be possible to postpone the publication of the book

> – until she is dead and no longer capable of being affected by it. It is for purely humanitarian grounds I appeal to you; I haven't read the report and therefore cannot argue with its contents.[11]

Clearly, however, Dingwall's animosity towards Price remained strong enough not to act on Tabori's advice. In fact, he rather ghoulishly pasted the letter in his scrapbook among clippings of reviews and notices in newspapers about the publication and revelations of *The Haunting of Borley Rectory*.

Despite all this, however, Price's investigation far outlived his rivals' criticism, and the site on which Borley Rectory stood is now a popular pilgrimage location for today's ghost hunters.

In 1948, psychical researcher B. Abdy Collins decided to revisit Miss Morton's written account of living in a haunted house. Perhaps he had been inspired by Price's recent work and wanted a Borley legacy of his own. Moreover, Price's approach to investigation was innovative despite its many flaws, so it seems natural that other writers and researchers would want to have a go at this new style of ghost hunting. Collins, in any case, found Miss Morton's report disappointing and full of holes. There was no ephemera or additional notes held at the SPR archives; what was published in the *Proceedings* over 60 years prior was all the evidence that existed of Morton's experiences. He believed it was time for a more thorough investigation.

While the society kept the details of the house confidential, like ghosts themselves these things have ways of being revealed. Collins' attention was drawn back to the case after the location was proven to be a particular house in Cheltenham situated on Pittville Circus Road.[12] With a little digging of my own, I found the more recent history of the property in question. Called 'St Anne's', it has since been converted into 16 high-end apartments, but a few years ago was up for commercial sale in a similar state of disrepair as it was when Mr S. died. While now the flats are decked out in gleaming art deco style, the photographs inside before it was sold and modernised really gave it a haunted vibe. It was dark, drab and in places rather cramped and tight for such a large dwelling. Mouldy curtains trailed on even mouldier carpet, too long for the windows from which they hung. In one photograph, a radiator appears to have been smouldering dangerously for several years, two teardrop-shaped scorch marks licking up the walls above it.

Upon learning the true location of the house, Collins wasted no time in conducting his own investigation. He travelled to Cheltenham in March 1946, armed with a camera and notebook, and the superintendent of the building gave him free rein to investigate the house and garden. It's clear from his recollection of this encounter with the superintendent that he had always intended to get a book out of his visit to the Mortons' old house. Collins sounded a little miffed, describing it as 'so ordinary and modern'; he was, perhaps, hoping for the rot and squalor, the damp and the frog infestation, that Price found at Borley. Perhaps if it had looked in the same sorry shape as I saw it when it was recently up for sale, he'd have been much happier. As it was, though, he could hardly understand how the property ever gained its reputation for being haunted.

Now certain of the report's location, Collins was able to clarify some of the other obscured details. The Morton surname, it turned out, was a pseudonym. The writer of the report, Miss Morton, was really named Rose Despard, daughter of Captain Despard. Mr S. was revealed to be a man named Henry Swinhoe, and his second wife, the alleged revenant, was Imogen Swinhoe. Her death certificate gave the official cause of her demise as 'Dipsomania: 6 months'; dipsomania was a commonly used term in the nineteenth century to mean a variety of medical problems caused by alcohol addiction. A single photograph of Imogen can be found online, and, if she really *was* the ghost, makes the calmness of Rose Despard's investigation even more baffling. Imogen was, in short, a terrifying woman even in life. She looks small, yet formidable; swamped in black mourning dress, she's wearing a large, billowing veil that scrapes her hair back. Two shrewd eyes look at some object in the corner, and her mouth is slightly twisted, giving her a rather cruel, disdainful expression. If I'd met her ghost in the dark, with my stub of candle guttering out, I would probably have screamed very, very loudly.

Collins' own tour of the house appeared to have been fairly fruitless; it's clear that Imogen's ghost, if it existed at all, no longer troubled the occupiers of St Anne's house. He instead casts a sceptical eye over Rose Despard's report, criticising the missteps and bungling of evidence from a more modern perspective of psychical research and more formalised data gathering. One of his key points targets Rose's description of the three sightings independent of her own. Her sister, the Despard's housekeeper and her little brother all saw figures that Rose believed matched her own encounter with Imogen's ghost. What Collins rightly draws attention to is the fact that all three of them first 'mistake' the spirit for a living person. Rose's sister believed she'd seen a nun, the housekeeper an intruder, and her brother thought it was a visitor to the house; not one of them reacted with the assumption that they'd seen a ghost.

Despite haunted house cases being visited and revisited, they have never fallen out of favour with ghost hunters. It is the quintessential fieldwork of the job, after all, and has become a rather lucrative business, too. While the Society for Psychical Research began more frequently to concern itself with telepathy and psychic forces as the twenty-first century approached, Peter Underwood made haunted houses a primary focus of his investigations with the Ghost Club Society (a bitter rift with the original Ghost Club had led him to create his own separate version, with some members defecting along with him). Gone were the days of sitting around in a members' club sharing spooky stories; Underwood had his new team tackling ghosts on their home turf. His 1994 book, *Nights in Haunted Houses*, demonstrates the breadth of investigations he undertook, each different to the last in terms of phenomena and experiences. Just as Collins went back to the Cheltenham house, Underwood decided it was time the Ghost Club Society faced the Moby Dick of haunted houses: Borley Rectory. Three weeks prior to Underwood's arrival, the rectory, now nothing more than charred ruins, had changed hands once again to James and Cathy Turner, and while Underwood and colleague Thomas Brown were the first ghost hunters to investigate the site during the Turner's ownership, within three years the 'continual harassment and inconvenience of unannounced visitors at all times of the day and night' (one ghost-hunting group even included someone carrying a shotgun, ready to fire at the ghostly nun) proved too much for them and they moved away from the site.[13] But it wasn't just their annoyance with amateur ghost hunters that drove them out. Cathy, while sitting in the ruins of the rectory reading a book, heard phantom footsteps that sounded as though they were coming from inside the building that no longer existed, as though they were walking on floorboards which had now disintegrated into blackened dust.

At the time of Underwood's investigations, ghost hunting was becoming much more equipment-based (although he always emphasised the need to keep things simple). Parapsychology was taking precedence over the techniques used by Spiritualists and psychical researchers, and a much more scientific, almost geological, approach was brought to the investigation of haunted houses. It was no longer about detective work, staking out the locations as in a film noir, or making discoveries like the celebrated archaeologists of the early twentieth century.

One of the most popular modern theories regarding why buildings gain a reputation for being haunted is the notion that the natural frequencies of bricks, wood and stone are imperceptible to the human ear as sound but are still subtly felt in the body. If you've ever stood somewhere near a very low, rumbling noise – say a bus engine or construction work, or too near a speaker with powerful bass – you've probably noticed the way it seems to stir up your stomach in a nauseating manner. But, in this instance, you can hear the sound and therefore you're aware of what's causing your discomfort. In a haunted house, so say parapsychologists, the same thing is happening but on a much lower, less noticeable scale. The sound cannot be heard but still creates strange sensations that you can't quite put your finger on – weakness, tingling, feelings of dread with no apparent cause – and have no idea why they've suddenly come over you. This theory is not just about the low frequency; the low frequencies make us feel eerie and uncomfortable, but it's our own overactive imaginations that elevate the sensations and create a spooky reason for them. This is likely why people often report 'not liking' a particular room in a house, thinking that there's some sort of malevolent spirit waiting within.

In 2004, a team of parapsychologists tested the theory that 'haunted' buildings do emit remarkable frequencies, at Muncaster Castle in Cumbria. The Tapestry Room was renowned for being haunted, so they conducted an investigation overnight with sensors designed to measure even the tiniest fluctuations in the environment's magnetic field. Reports of strange experiences had been especially prevalent on and around the bed, so a sensor was placed either side and on the pillow, while a baseline sensor was put in the approximate midpoint of the room. What they found was that the sensor on the bed pillow recorded significantly larger fluctuations in the magnetic field than the baseline sensor in the middle of the room, with an overall range of 195 nanotesla (nT) compared to 131nT.[14] The 'bursts' were much more frequent and strong by the pillow, too, and were actually very similar to the level of variation synthetically created in laboratory experiments that tested the effects of magnetic fields on bodily sensations. They concluded that in reputably haunted buildings where such magnetic field variation existed similar to what they recorded in Muncaster Castle's Tapestry Room, combined with a level of suggestibility and expectation from the percipient, reports of witnessing paranormal activity would be highly likely.

Thinking back to my experience of the ASSAP training day, magnetic field variations throughout the thick stone walls of the Old Prison may well have explained the way everyone's experiences were purely sensation-based. However, this theory only accounts for sensations – the 'feeling' that a house is haunted. It doesn't account for spectres witnessed by multiple people, knocks and bangs, disembodied noises, and the movement and materialisation of objects. Haunted houses are simply too multifaceted, each its own unique case, for every ghost to be explained through magnetic field anomalies.

CHAPTER FOUR

The Ghost in the Stereoscope

The Adelphi Hotel in Liverpool is known as one of the most haunted buildings in the United Kingdom. It is, like all renowned spooky hotspots, a magnet for ghost hunters, but it rose to particular prominence in July 2021 after it was visited by Lee and Linzi Steer, a formerly married ghost-hunting couple from Yorkshire.[1] Among other phenomena such as scratching sounds and eerie whistling noises, an image was captured as they live-streamed their investigation to their 1.8 million Facebook followers. It shows a blurry view of the end of a corridor with salmon-pink walls and clashing dark carpet. In the left-hand corner stand two small, thin figures. They don't quite reach the dado rail midway up the wall, and appear to be holding hands. Local news sites leapt on the story, pointing out the uncanny resemblance between the phantom figures in the photograph and the twin girls in Stanley Kubrick's adaptation of Stephen King's haunted-hotel novel *The Shining*.

It's a rather effective, unsettling picture – until you see the same angle of the corridor with a sharper focus. The pair of ghostly children, it turned out, were two fire extinguishers fixed to the wall.

This is but a recent example in the long history of using cameras, photography and recording equipment to capture, and manipulate, evidence of paranormal activity. When the Fox sisters began their séances, they were referred to as the 'spiritual telegraph'. With proponents of Spiritualism and ghost hunting seeking to legitimise and defend themselves against their critics, the two have always been tied to the latest advances in technology. As new technologies began to be developed in the late nineteenth century, they fundamentally changed what some people thought was physically possible. The wireless network allowed instant communication across enormous distances, and photography meant that moments in time, and people, could be captured with astonishing clarity. Technology was fuel for the Spiritualist movement, members of which claimed that such advances seemed like fantasy not too long ago – and yet were now fact. If telegraphic messages and photographs were now a reality, who could say that spiritual phenomena wouldn't also be proven in the years to come?

It wasn't simply to help their argument that Spiritualists and psychical researchers were so fascinated by new technologies. Prior to these recent inventions, séances and experiments had relied upon witness testimony, diagrams and illustrations – all of which were subject to the bias, misremembrances and fraudulent intentions of those who created the reports. Séances had lacked objective evidence, and so as photographic techniques developed and equipment became more accessible to the layperson, no Spiritualistic sitting could be considered serious without the presence of a camera.

In the first decades of the twentieth century, photography became particularly useful when a new trend emerged among mediums: materialisation. Hidden in a séance cabinet, the medium would writhe and groan before dramatically – but briefly – opening the door or curtains to reveal a paranormal manifestation in the form of squidgy white ectoplasm or a full-bodied spirit draped in glowing cloth. A photograph timed at the right moment could capture the fleeting glimpse of the spirit forever, which could then be analysed and interpreted at the researcher's leisure.

But such a seemingly clear answer to the question of evidence was more problematic than first thought.

Spirit photography was first designed as a joke, an entertaining way to trick the Victorian public, who loved a good optical illusion, and to show the sorts of marvellous feats of which this new technology was capable. In 1854, the London Stereoscopic Company was formed in order to demonstrate the novel ways in which photographs could be used to create surprising and shocking scenes. A stereoscope is a viewing device, shaped like a very clunky pair of glasses, which, when used to look at two almost identical images side by side, creates a three-dimensional image. Through them, the paying Victorian public could enter a new reality, and the company's most popular stereoscopic photographs were those that depicted theatrical scenes. Headed by George Swan Nottage, the London Stereoscopic Company achieved great success, and one of the their series of cards was called 'The Ghost in the Stereoscope'.[2] These showed translucent, skeletal ghosts bursting in on unsuspecting victims. They're still quite creepy to look at today; in one of them, a sepia ghost, bony hands outstretched, seems to reach for two horrified men sitting at a table – but the ghost's eyes are fixed on some point in the middle distance, above the men. You can see a colourised red chair through the cloth draped over the ghost, and the nearer to the floor we look, the less visible the ghost becomes. It's very effective – and it's also very fake. But it was never meant to be taken as a real photograph; it was an early form of special effects, a way of demonstrating the marvellous capabilities of image manipulation to create an illusion of reality.

'The Ghost in the Stereoscope', however, evolved quite the legacy. Soon, photographers were recreating the images but under the pretext that they *were* real, that what they captured on the photographic plate was what was really there at the time. Spirit photographers opened for business across the world. In 1872, Frederick Hudson, in his small studio in London, took what was thought to be the first spirit photograph in Britain. During a table-tilting séance in Hudson's studio with well-known medium Elisabeth Guppy and her husband, the visiting spirit told the three of them to take a photograph. Mr Guppy sat before the camera while Elisabeth stayed in a dark cabinet; when the photograph was developed, Mr Guppy appeared to have been accompanied by a large ghost standing by him.

By this time, spirit photography was also already well-established in the US, and no one in the business was perhaps more notorious than the American jewellery engraver William Mumler, who began snapping ghosts in the 1860s.

Mumler's photographs look like typical Victorian portraits, with a stoic-faced sitter in front of a plain background. Only, they're not alone; a pale, translucent face or figure hovers over them, sometimes overlapping with their shoulder or, if the phantom figure is a child or baby, positioned in the sitter's lap. His services weren't cheap, either; a session would cost $10, equivalent to nearly $400 (£315) today, while a handful of prints of his photographs sold for $5 each.

What's interesting about Mumler's photographs is that they seem so immediately risky because of his high-profile clients (and, indeed, high-profile ghosts). Emma Hardinge Britten for instance, whose involvement in Spiritualism included the list of rules to be followed for a successful séance, was one of Mumler's patrons. She sat for him on several occasions in order for him to experiment with his camera. One photograph depicts her enshrouded by the ghost of Beethoven. When Abraham Lincoln was assassinated in 1865, his widow Mary Todd Lincoln sought Mumler's services. The photograph obtained depicts Mary in her widow's garb sitting towards the left of the portrait, leaving room for Lincoln's ghost to stand behind her. He is looking pensively down at his wife, hands loosely draped on her shoulders. Yet there's something strangely flat and disjointed about his hands, as though they've been painted on or belong to another.

It was such suspicious details as this that brought a high-profile court case to Mumler's doorstep. In 1869, his photographs were the subject of a preliminary hearing in the city of New York for the Court of Special Sessions. He was acquitted of the charge of fraud due to a lack of sufficient evidence to prove that his photographs were fakes. It seems like an implausible outcome, but the trial wasn't exactly fair and impartial: many of the members of the jury were Spiritualists, and some of the evidence produced in favour of Mumler was given by thoroughly respectable individuals. Judge Edmonds, for example, who was also a senator for the state of New York, gave evidence to support Mumler. Edmonds was the author of a two-volume book on Spiritualism, and had previously sat for Mumler to receive a handful of spirit photographs.

While Mumler got off lightly, the debate around him still raged. This wasn't helped when Frederick Hudson was pulled apart by his peers in the *Spiritualist* journal for manipulating his photographs through double exposure in 1872, and similarly in 1875, when a renowned spirit photographer in France was successfully prosecuted for fraud. Édouard Buguet began producing spirit photographs for 20 francs a go (equivalent to around a week's wages[3]) in 1873 in his studio in Montmartre, and was soon advertised in the French Spiritualist magazine *Revue Spirite*. A police raid of his studio uncovered Buguet's tricks; dummies were found – both adult- and child-size – as well as fake beards, costume pieces and boxes full of hundreds of photographs of people of all ages and appearances (many of whom were still alive and well). According to Rolf H. Krauss, Buguet's receptionist would find out as much information as possible about the ghost whose presence was desired in the customer's photograph.[4] The information would be passed to Buguet who would pre-prepare a photographic plate using a likeness from his box of photographs pasted onto one of the dummies. He would then double expose this plate and use it to photograph the customer – the dummy faded and translucent in the background. This technique was especially effective for deceased persons who hadn't been photographed while alive, as the memory of their appearance would have faded in the customer's mind – thus making it easier for them to accept a likeness as the truth. Buguet was imprisoned for a year and fined 500 francs.

Still, however, the high-profile cases of photographers such as Buguet and Mumler didn't do much to cease the popularity of capturing ghosts on camera. By the end of the nineteenth century, Eleanor Sidgwick (who we first met in Chapter 2) was becoming irritated by the ease with which photographers were able to earn a tidy sum from producing fraudulent images of ghosts hovering around their loved ones. Despite the technical skill involved in chemical manipulation and double exposures, Sidgwick called spirit photography 'practically conjuring' in her paper delivered to the Society for Psychical Research.[5] She was an eminent physicist and probably understood more about photography than most people, but she wrote in her paper that the reason this phenomenon was so successful at duping people was because the processes of development were obscure and complicated to the general public, not to mention 'the number of ways in which sham ghost pictures may be done'. Trickery within a séance room was fairly limited, but a photographer alone in their darkroom was free to use a dozen different techniques to produce the ghostly 'extra' in the portrait.

Nevertheless, spirit photographs seemed to remain relatively untarnished even as frauds were exposed. This is a recurring trend with Spiritualism and ghost hunting, as we'll continue to see; no matter how many people are proven to fake their evidence, belief in ghosts remains unwavering. Decades after Mumler's death, his infamous court case was still being debated – and because he had been acquitted, he became something of a parable for Spiritualists. For example, in 1911, James Coates published *Photographing the Invisible*, which weighed in on Buguet and Mumler nearly 50 years after the fact. Of Buguet, Coates described him as the 'fly in the ointment; the counterfeit among the coins [...] which helped the man in the armchair who has never experimented, to say, "All the ointment is bad, all the coins are counterfeit; all your psycho-physical phenomena are fraudulent, and all your mediums are imposters."'[6] Or, indeed, the woman in the armchair, since Coates repeatedly went after Eleanor Sidgwick. 'Actual experience and reliable testimony of competent experts', he wrote, 'go for nothing because this clever woman assumes fraud.' He dug up old debates and took aim at investigators now well into their old age; he sent a packet of psychic photographs to Sir Oliver Lodge, encouraging him to investigate, but never heard back. He also reworked Emma Hardinge Britten's séance rules for the scientific twentieth century, confident that spirit photography would be the main focus of Spiritualists: 'extras', he declared, were 'crystallisations of thought', and the best photographs were produced by cameras which had long been used by a psychically sensitive photographer who, like a cup becoming brown with tea stains, had imbued their device with their magnetic 'nervaura'.

But the twentieth century, largely, had a different use for spirit photography. Individual portraits, for the most part, ceased to be in vogue; instead, the camera was employed in the séance room itself to document the process through a seemingly objective lens. In 1909, the German physician Albert von Schrenck-Notzing began a series of sittings with a young French woman whose feats of materialisation were supposedly remarkable. Schrenck-Notzing had been fascinated by Spiritualistic phenomena for many years, and was particularly struck by the rise and fall of notorious medium Eusapia Palladino, whom he once caught making 'spirit messages' appear on his shirt cuff with a tiny pencil nub hidden in her hand. Palladino's exposure hadn't put him off the pursuit for evidence, however; he was still convinced that there *was* truth in spirit manifestation, but he needed a more reliable medium and a more reliable way of documenting evidence. Photography, he thought, would allow him to discern fact from fakery and, in doing so, would finally prove the existence of spirits. Echoing many Spiritualists at the time, Schrenck-Notzing lists 'colour photography', 'wireless telegraphy' and 'radio-activity'[7] as former supposed impossibilities of nature; spirit manifestation was clearly the next frontier of scientific discovery.

The sittings with Eva Carrière, spanning four years, were published in the hefty volume *Phenomena of Materialisation*, which was translated into English in 1923. Eva C., as she's called in the book, was 23 years old when she began to sit for Schrenck-Notzing. She was described by him as having a 'vivid imagination, which is sometimes so exaggerated that truth and fiction can no longer be distinguished', and that she acted as though her 'feminine charms' had more effect on men than they did. Throughout the book, Schrenck-Notzing painted her as someone who was moody, irritable, childish and prone to accusation and jumping to conclusions. He paid a great deal of attention to Eva C.'s personality, and, to his discredit, less to that of the older woman who never let Eva out of her sight: the French sculptor Juliette Bisson. He said that Eva's attitude towards Madame Bisson was like 'that of a faithful dog to its master' and frequently called Bisson her 'protectress'. Bisson sewed Eva into clothes made especially for the sittings, allegedly checked Eva's body for concealed articles, gripped her hands and feet, and held her when she was in the painful grip of mediumistic labour. Yet he never seemed to question this dynamic, especially when, as we will see, evidence cast a doubtful eye over Bisson's role in the phenomena.

Beginning in 1909, the sittings followed a similar and repetitive pattern throughout. A number of cameras were set up, poised at the séance cabinet, and lighting was used in varying combinations. Eva was seated inside the cabinet and put into a trance; the curtains were closed. She began to moan and writhe, and the curtains jolted and flapped. Suddenly, they were pulled apart and Eva briefly showed the assembled group – sometimes just Schrenck-Notzing and Madame Bisson, but sometimes others who joined them – the materialised spirit. If the person operating the camera was quick enough, they could take a photograph before the curtains were closed again. Schrenck-Notzing wrote up a report of each sitting, attaching diagrams of the séance room as well as including a mixture of photographs and artist illustrations of the manifestation.

Photography in spiritualistic sittings was used with the idea that it would diminish the scepticism surrounding purely narrative accounts, but it is the visual material of Schrenck-Notzing's book that damns the otherwise eerie descriptions of the phenomena. The 'teleplastic' structures seeming to emanate from Eva's body were presented through words as strange, shifting, formless and glowing, appearing and disappearing in an unnatural manner. For example, of the sitting on 25 November 1909, Schrenck-Notzing wrote of an 'illumination of the curtain' which 'resembled a bright phosphorescent strip', morphing and growing into a shapeless cloud that popped in and out of vision like a flashing bulb. It trembled 'with a fluctuating motion, then it became visibly brighter and more solid, until it changed into a white luminous material' and finally fizzled out of existence in front of the onlookers. Multiple sittings described similar phenomena, and the imagery conjured up etheric flotsam that was unlike any kind of matter we're used to perceiving – words seemed to be incapable of truly capturing the appearance and movement of these paranormal projections.

When you turn the page to the accompanying photographs, however, the disappointment is palpable. One after the other, images of Eva mid-materialisation are presented, all with similar results: she sits grimacing in a chair, curtains scrunched in her hands as a strip of paper is draped over her shoulder or a ragged cord of white muslin dangles from her clenched knees. Occasionally, illustrations of the scene are presented as well, complicating the matter further. It shows, I think, just how much we embellish what we believe we've witnessed; if we're convinced that something was paranormal, then our retellings and depictions that come from memory will always take on a paranormal tinge. These illustrations present a creepy formlessness and glowing quality to the phenomena issuing from Eva. The teleplasm in the photographs, on the other hand, falls flat in comparison; there is none of the phosphorescence Schrenck-Notzing talks about, and you can easily pick out creases, seams and rents in the material that do nothing to enhance the ethereal quality of what we're supposed to be looking at.

As with so many of the cases we've come across already, we have to wonder why Schrenck-Notzing didn't see the fraudulent nature of Eva's materialisations. It can't be down to cultural ideas of mediums or the relative novelty of photographs because others around him were equally suspicious. A curious incident related in the foreword of this book describes how someone, who employed the services of a private detective agency to follow Eva and Madame Bisson, also managed to acquire a series of photographic negatives from Schrenck-Notzing's collection. These were presumably stolen by one of the numerous collaborators and physicians he invited to watch Eva C. in action (including the Irish poet W. B. Yeats). It was not revealed who hired the detectives, but Schrenck-Notzing must have come in contact with them because he recalled how they were unable to find 'any proof of fraud'.

The prospect that Eva C. was lying to him, however, was always on Schrenck-Notzing's mind, and appeared to become even more of a pressing issue as the sittings developed. One common thread running throughout the history of ghost hunting is the constant need to show increasingly impressive displays of spooky phenomena, and the case of Eva C. was no exception.

After the death of Monsieur Bisson, Eva's materialisation started to take on a new form. Previously, her teleplasm had formed appendages: hands, fingers, stubby limbs. Now, as the curtains parted, facial features could be discerned among the shreds and scraps of material. Photographs revealed faces amid a crumpled mass attached to Eva's hair, and the bearded head of Madame Bisson's late husband peeped from behind the curtain. All of them bore the same paper-like appearance; creased, smudged and flat.

Then came an array of suspicious discoveries. On 21 August 1912, a sceptical observer, Dr A., pointed out a series of small, new pinholes in the cabinet curtains. The holes came in pairs, exactly as though something had been attached to the fabric using a safety pin. When the positions of these holes were compared to the photographs, they matched with places where materialised faces and forms poked out from behind the curtains. Nine days later, a number of white crumbs and pea-sized balls were found on the floor of the cabinet; microscopic analysis showed these particles to contain wood fibre – they were bits of paper. Most damning of all was the sitting on 27 November, in which a face was materialised along with letters hidden amid the creases. On further inspection, both the letters and female face appeared to match the masthead of French newspaper *Le Miroir.* Schrenck-Notzing ended his description of this sitting with a rather telling sentence: 'I cannot form any opinion on this curious result.'

Confronted with these increasingly dodgy events, Schrenck-Notzing began to make elaborate excuses for Eva C. We saw the precursor to this in the opening of his book, when he told his readers that even though Eusapia Palladino was proven to be fraudulent, it may simply have been that she acted while still in her mediumistic trance – it didn't prove that she was knowingly fraudulent. The same line of thought extended now to Eva C. In the case of the pinholes, Schrenck-Notzing said there was nothing wrong with this: the teleplasm was ethereal and fleeting, and while Eva should have declared that she had a stash of pins, she was merely helping the investigators with their research by pinning the materialised matter to the curtain. Of the balls of paper, he explained that they may have been brought in on the sole of someone's shoe. And in relation to *Le Miroir*, he offered a continuation of his theory as to why the materialised faces looked like flat, unmoving images on paper: they weren't spirits themselves, but two-dimensional projections onto a flattened teleplasmic screen. Eva C. endorsed this, and told him that she *meant* to materialise the letters of 'miroir' not in relation to the French newspaper, but in terms of the word itself: it was a textual message from beyond the grave.

While there were moments of doubt throughout *Phenomena of Materialisation*, there seemed to be a sense from Schrenck-Notzing that nothing would shake his belief. His explanations were as creative as they were limitless, and even when Madame Bisson sent in her own reports of the private sittings between herself and Eva, he didn't for a second question the veracity of remarkable phenomena he did not himself witness.

This is perhaps the place to turn the focus to the dubious Madame Bisson. We've seen how peculiar *Phenomena of Materialisation* is as a book investigating psychical research and spirit manifestation, demonstrating plainly the split between events recounted by subjective memory and events captured by an objective photograph. But there's another aspect of the text which makes it, at times, deeply uncomfortable to read. We've already explored in previous chapters how sexualised séances could be, and that normal rules of social etiquette didn't seem to apply among circles, whose participants touched and probed each other's bodies in the name of Spiritualistic evidence. Schrenck-Notzing's experiments with Eva C. are some of the most troubling in relation to this, but not just in terms of what the researcher does to the medium. Madame Bisson is the true ghoul of this book, lurking in the background and always ready to photograph Eva naked and poke her fingers into her genitalia (which, horribly, Schrenck-Notzing also did on the rare occasion Bisson was absent from a sitting – much to Eva's distress). Eva lived in a room connected to Madame Bisson's art studio, and was hardly ever out of the older woman's sight. The photographs Madame Bisson sent to Schrenck-Notzing of her private séances with Eva are, quite frankly, pornographic. Out of context, Eva's expressions might be mistaken as showing sexual pleasure rather than the physical efforts of materialising spirits. Throughout his report, Schrenck-Notzing painted Eva as an emotionally volatile young woman, but never did he question the way Bisson treated her, nor did he ever seem remorseful of the things he himself put the young medium through. Of all the morally objectionable incidents in the history of ghost hunting, this is one of the most problematic – and it was by no means the last time exploitation of this kind happened to a medium under the guise of investigation.

At around the same time as the peculiar experiments with Eva C. were taking place in France, another scientist was conducting photographic investigations of his own in Ireland. William Jackson Crawford was an esteemed lecturer in mechanical engineering at Queen's University, Belfast, and had written influential texts and undergraduate study guides on the subject such as *Calculations on the Entropy-temperature Chart* (1912). After he began to investigate Spiritualism, however, his interests changed and he instead focused his writing on his experiments with the young Irish medium Kathleen Goligher.

In 1921, a year after Crawford's unexpected death by suicide, his comprehensive photographic study of Kathleen's phenomena was published under the title *The Psychic Structures at the Goligher Circle*. He had positioned the text as a scientific one, no different from his *Calculations* in terms of technical rigour. Kathleen's materialisation was unlike Eva C.'s teleplasm in many ways; where Eva's was so weak and gelatinous it had to be pinned to the curtain, Kathleen's was a long, phallic limb that wrapped around table legs like a snake, left deep impressions in soft clay, and was even capable of gripping and lifting up tables and chairs. In order to emphasise that he was approaching the investigation from his position as a mechanical engineer, he used the same terminology to describe the phenomena, calling the materialisations 'cantilevers' and rods capable of exerting tremendous force.[8] The equipment employed, too, demonstrates his need for this work to be treated as equal to his academic research. Indeed, if you flick quickly through the book, you come across diagrams and charts that wouldn't be out of place on a classroom board during a lesson on Newton's laws of motion. You can even take certain sections of the book out of context and you wouldn't be able to tell that it was actually about a séance. For example, he wrote that 'the gripping part of the structure must have special labour expended on it before it is able to perform its function. It must, of necessity, be a differentiated portion of the structure. It follows from the point of view of the saving of energy that the fewer the number of these differentiated portions the better.'

After first witnessing Kathleen's powers, Crawford quickly set up sittings – hundreds of them, in fact, over a period of years – so that he could photograph the evidence. The first photograph included is, actually, quite spooky. We see a group gathered around a table, a white translucent tendril about as wide as an arm points up towards the ceiling, its tip curled like the frond of a fern. Besides suspecting the usual double exposure and chemical tricks, there's nothing here that smacks of fraud. It's certainly more impressive than the Schrenck-Notzing photographs. However, the rest of the text includes photographs, and discussion of photography, which hint at the true nature of Kathleen's phantom cantilevers.

Crawford described early on in the book that the so-called 'psychic structures' were invisible. Not only that, but any attempts to illuminate the room with anything stronger than a feeble red light bulb appeared to harm Kathleen. The spirits who produced the ectoplasmic rods, termed 'operators' by Crawford, thus made it clear that any attempts to use flash photography would seriously injure their medium. This was proven when Crawford attempted to take a picture of Kathleen mid-materialisation:

> ...when the flash occurred she trembled violently and her body jerked about spasmodically for ten minutes or longer. If the structure shown, or one like it, had been below the surface of the table levitating it, the structure would have been under considerable mechanical stress, much energy from the medium's body would have been required to organise it, and the resulting disturbance to the medium when the flash occurred would have been much greater. For this reason the operators never allowed us to take a flashlight photograph of the levitated table. The disturbance to the bodily organism of the medium would have been too great, and might possibly have been dangerous.

This is something that formed part of the wider mythology of mediumistic phenomena once cameras became widely used in séances: the spirits were acutely sensitive to light, and could only stand the dimmest of red lamps or thin candles, and any attempts to use flash photography – without first receiving clear permission from the medium or the spirits – would injure their ectoplasmic forms and possibly prevent any further activity in the séance room. Investigators took this seriously, not wishing to lose a medium whose trust and willingness to participate in humiliating experiments had been hard won.

Crawford's photographs were thus only taken when Kathleen's 'operators' allowed it, when the set-up was perfect, advance warning was given and the means to conceal fraudulent evidence was easily obtained. While the initial photograph in *The Psychic Structures of the Goligher Circle* is eerie, the rest demonstrate the same flatness, the same cheesecloth appearance, as Schrenck-Notzing's images of Eva C.'s phenomena. And, strangely, Crawford encountered a very similar suspicious anomaly in Kathleen, which, like Schrenck-Notzing, he went to great lengths to explain using Spiritualistic theories.

As he established that the ectoplasmic cantilevers were invisible unless photographed when allowed, Crawford set up a different way to capture their image. He filled a shallow tin with a smooth, flat wad of putty and placed it beneath the séance table, and asked the operators to press the end of a psychic rod into the substance. Once impressed, Crawford made plaster casts and photographed the results. It is difficult not to be instantly sceptical, which Crawford pre-empts and says, 'to a careless eye, or to an eye which has not seen any other impressions taken at the circle, [it] looks more or less like the mark of the sole of a lady's shoe.' I do, alas, apparently have a careless eye. It's hard to be persuaded by Crawford, especially in the next set of photographs taken of impressions in which the criss-cross pattern of a stocking puckered the surface of the putty. And like the tiny balls of paper found in Eva C.'s cabinet, the putty was covered in coal dust, grit and crumbs – as one might expect to find on the bottom of someone's foot. But Crawford had answers to these conundrums: the operators created the rods *from* the end of Kathleen's foot, whether she wore a stocking or a shoe or nothing at all, and, being of a 'glutinous, fibrous nature', took with them any fragments of debris or patterns of woven stocking fabric as they moved towards the putty.

Photographs of Kathleen's stockinged feet then proliferate the text; the sole of the wrinkled, over-used fabric is shown to have smears of clay that coincide with the marks left behind in the putty. Surely this would be enough for Crawford to suspect foul play, but, again, he created an elaborate explanation. The psychic rod 'retired into the body of the medium from between her shoes', he claimed, 'through the space at the instep, up the side of the shoe, down the inside and thence to the foot.' In other words, it snaked out of her shoe, touched the wet putty, and then retreated again back through her stocking and left behind the white residue. But it was not just Kathleen who seemed to have a hand in the fraudulent phenomena. During one of these putty experiments, a friend of the Golighers and regular participant in the séances, Mr Morrison, exclaimed that his boot was touched by a psychic rod. A deep impression was made in the tin, and clay was found on the surface of Morrison's shoe.

Why couldn't Crawford see what is painfully obvious to anyone who looks at these photographs?

As with the Schrenck-Notzing experiments, and so many other séances we have already seen, there was also a troubling sexual undercurrent throughout Crawford's investigations. He was fascinated by Kathleen's body, by the way she lolled and moaned in her chair, by the thought of phallic rods protruding out of and retreating back into her flesh. Towards the end of the book, he describes how he 'felt the medium's breasts during the occurrence of psychic action. They became very hard and full.' All in the name of science, of course.

Moreover, there was a great deal at stake for Crawford in terms of his profession; as a renowned engineer, he had much more of a reputation to uphold among his rational, scientific community than Schrenck-Notzing. This was why he tried so hard to legitimise his experiments and the resulting photographs as worthy of attention within his academic community. But what was noticeable to many was the way in which Crawford had been clearly duped, not helped by his infatuation with Kathleen, and the book was slammed by critics within both engineering and psychical research circles. Crawford was barely cold in his grave while prominent figures such as the surgeon Charles Marsh Beadnell and psychical researcher Hereward Carrington lambasted the experiments as demonstrating nothing more than Crawford's own inability to see when he was being hoodwinked. It's not known whether the toll of these experiments, and of the growing realisation that Kathleen had been exploiting him, contributed to his suicide, although Eric Dingwall declared that Crawford had confessed as much to him before his death.

As the twentieth century progressed, psychical researchers became less interested in trying to capture visual evidence of ghosts. Experiments in the 1960s hailed a new form of investigation that sought to use digital equipment to bridge the gap between the living and the dead, and it involved using radio and audio-recording devices to capture paranormal voices. The Latvian parapsychologist and thinker Konstantin Raudive published an influential text, bombastically titled *Breakthrough* (1971), which described the techniques by which the dead could make themselves heard once more.

Raudive explained that this was a purely scientific endeavour. It involved, he said, acoustics, and 'leads to empirically provable reality with a factual background.'[9] His friend, a wealthy man named Friedrich Jürgenson, had been dabbling with a tape recorder, and had found something very strange indeed. He played his samples to Mrs Jürgenson, Raudive and Raudive's wife, Latvian essayist Dr Zenta Maurina. Raudive could immediately hear eerie voices over the static of the recording, but, he explained, 'our unpractised ears had great difficulty in identifying them.' The way he wrote about how the group's sense of hearing 'could gear itself' to the words of the dead makes the process sound like eyes adjusting to darkness, gradually picking out patterns that reveal themselves from the gloom.

Intrigued, the group decided to make their own recording together, and again could hear human voices muttering under the radio's fizz and crackle. Not only that, but they found that whoever was on the other side was aware of the group, and could hear them talking. At one point, Dr Maurina says that the voices sound happy. Immediately, a voice answers, 'Nonsense!'

As a psychologist, Raudive's instinct was to explore the voices as a manifestation of the unconscious mind, and as a scientist he sought to explain the voices coming through the radio static as 'coincidental sound-freaks from transmitting stations'. While the latter seemed like a valid conclusion, clearly it didn't answer Raudive's questions to his satisfaction. He was hooked, and turned Jürgenson's tinkering into a full-blown laboratory experiment spanning five years and tens of thousands of samples. It took him three months to perfect the art of capturing voices, a technique that involved running the radio and microphone simultaneously through the recorder. This way, Raudive's own questions and comments were recorded as well as the alleged ethereal answers.

His methods became more scientific, as though by complicating the process and employing more equipment the recordings were further legitimised. On his technique of using diode recordings, he explained that the aerial

(6–8mm) has to be positioned to a precise specification in order to capture the ghostly voices as clearly as possible. This way, Raudive explained, the dead would speak 'directly onto the tape; they have a spaceless quality, an immediate impact and their diction is remarkably clear; they are instantly received and can be heard without atmospheric interferences.' Despite sounding clear, Raudive considered the eccentricities of the samples. From his analysis, the voices spoke quickly, with garbled syllables, and often switched from one European language to another in the same sentence. His answer to this was that communication in the afterlife must be very different from our own; the dead don't speak another language, per se, but in a different style that compresses, truncates, and relies heavily on slang and words borrowed from other countries. As we will see in the next chapter, if the afterlife defies boundaries and all spirits begin their next journey on equal footing, then surely all the world's languages would be one interchangeable mode of communication. But because of Raudive's Latvian tongue and collection of knowledge of other European languages, the spirits limited themselves to words he could understand.

After explaining his methods and defending the scientific credibility behind the experiments, Raudive began to analyse a selection of samples. This is where *Breakthrough* gets very odd. An early example in the text is one which Raudive thought came from his late aunt, whom he believed popped up now and again in the recordings to give him a warning or other maudlin message in a mix of Latvian and German. The translation of her message is particularly uncanny: 'This is your aunt, dead, dead aunt. This is your aunt dead aunt. You godless, sing!'

The recordings soon took an even stranger turn. In the height of Spiritualism's table-tilting séances, famous ghosts such as Shakespeare would grace random families in their sitting rooms with haphazard lines of bad iambic pentameter.

Breakthrough demonstrates that this trend was far from over even in the 1960s. Tolstoy and Dostoevsky appear, followed by Carl Jung and Winston Churchill. Oliver Lodge, whose involvement in Spiritualism coincided with his influential work on radio transmission, was clearly admired by Raudive – it's no surprise, then, that he dropped in to say hello.

Dictators, too, began to make an appearance. Adolf Hitler, in particular, cropped up alarmingly often in Raudive's analyses of the voices. He responded to these dictators from a psychoanalytic angle, arguing that they are 'caught in a spiritual impasse from which, even in death, they cannot escape.' When Lenin's voice was recorded, he sounded like 'a patient in a lunatic asylum' whose words repeated a single phrase or idea over and over again. Hitler, too, sounded different to the other voices in the samples: 'he shows exactly the same traits that characterised him on earth: self-gratification (megalomania), persistence in pushing himself forward and a certain spiritual depravity – all sharply rejected by some of the other voice-entities.' Raudive gave as one example: 'Greetings, first you must have a sense of humour. Here is Major Hitler. Honour Hitler a little!'

Interestingly, Hitler spoke to Raudive in Latvian. You would think that even with the theory that the afterlife's languages fluidly intermingle, Hitler would still speak solely in German. For a psychoanalyst who learnt under Jung, it's peculiar that Raudive didn't approach this particular manifestation from another perspective – one that questions why he heard Lenin and Hitler at all. Latvia suffered periods of oppressive occupation in the Second World War: first the Soviet Union took control in 1940, then Hitler's Nazi regime seized the country between 1941 and 1943 before the Soviets resumed occupation in 1944. Following this, Raudive ended up in Sweden, exiled from his home country. He was directly affected by dictatorships; it must have been on his mind.

For all the grief we've already seen surrounding Spiritualism and psychical research, it's understandable that he would hear his relatives speaking to him through the recordings. And, as mentioned earlier, hearing 'celebrity' voices is a trend as old as séances themselves. Ghost hunting, I've found, is very much about yearning. It's a physical, tangible action towards something inherently metaphysical and intangible; it allows us to lessen the gap between the living and the dead, to bring us closer to the loved ones we've lost. What we don't see very much of, though, are investigators coming face-to-face with the ghosts of malevolent figures who haunted them even in life. Here, Raudive heard not only the voices of his dearly departed but of those responsible for his suffering. I suppose it's still about receiving closure which death so often denies us, but it comes from a different place. Hitler's cruelty was so massive, so widespread, that it must have created cognitive dissonance to be so individually persecuted by someone who didn't know you existed. By hearing Hitler's voice come through, speaking to Raudive in Latvian, perhaps he could then feel that Hitler *did* acknowledge him and *had* targeted him personally.

Speaking with ghosts allows us to soothe our immense grief, but, clearly, here it also allows us to confront those who hurt us without granting us closure. Not that Raudive admitted this. 'Tape-recorder, radio and microphone,' he wrote in the opening pages, 'give us facts in an entirely impersonal way and their objectivity cannot be challenged.' This is true, of course: we can't dispute that a tape recorder has picked up sounds. What *can* be challenged, however, is our interpretation of those sounds.

Raudive's text is still being examined by scientists keen to understand the way in which these phenomena were analysed. In 2015, for example, Michael A. Nees and Charlotte Phillips, based at Lafayette College in Pennsylvania, looked again at *Breakthrough* in order to understand whether the intention to find paranormal evidence makes us more likely to hear ghostly voices through white noise.[10] They were concerned by the fact that Raudive didn't consider human bias, namely pareidolia – the way our brains create recognisable patterns out of random stimuli – and how this can skew interpretations. They conducted an experiment with two groups of students, who were both given audio clips of electronic voice phenomena from popular paranormal reality TV show *Ghost Adventures* as well as samples of real human speech embedded in white noise. One group were told that the experiment was measuring people's ability to pick out voices in noisy environments; the other group were told that they were investigating the identification of paranormal voices in digital audio recordings.

Unsurprisingly, perhaps, they found that the group who were primed to listen out for paranormal voices believed they could identify far more words and phrases than those who were given the 'normal' context for the test. Their conclusion was that even the slightest mention of the paranormal was enough to demonstrably skew people's perception of stimuli to one that focused on ghostly interpretations. Most of us will have experienced this type of priming in our lives. As children, noises and shadows in the dark of a sleeping house were likely interpreted as having a spooky origin, because at that age, that's what we're primed to fear: monsters, ghosts, witches and aliens. As adults, if we hear a bang or creak, we first think of more realistic threats such as a burglar or a rat (unless, of course, we're staying in a house we're told is haunted).

What's interesting about Nees and Phillips' study is that the group who said they could hear paranormal voices interpreted those voices differently: no one heard the exact same words or phrases. Raudive's *Breakthrough* was published with an accompanying vinyl record so that readers could listen along to his samples. While the book itself is now rare – I had to consult it in the National Library of Wales' reading room – the recording, at the time of writing this, has been made publicly available on YouTube.[11] The narrator, Nadia Fowler, introduces each sample with a clipped, informative tone, announcing what you're *supposed* to hear before you hear it, forcing you to try to fit the phrase with the garbled sound. Often the voice speaks in a mixture of European languages, with phrases switching word-by-word from German to Latvian to English to French, and the narrator provides the translation. It's worth a listen, especially in relation to how Raudive described the voices in the text of *Breakthrough*; it further demonstrates what we witnessed with Schrenck-Notzing – textual descriptions of phenomena are creepy and strange, but the actual evidence itself never quite lives up to how it has been interpreted by the human imagination. The *Breakthrough* recordings do present voices: there are clear vocal words and sounds that aren't just white noise. However, it's rare that the phrase announced by the female narrator perfectly matches the recording. Voices are distorted by migraine-inducing static, but even so the interpretation seems to add syllables and sounds, distorting the sample to make it relate to Konstantin Raudive's name or a semi-logical answer to a question posed to the radio equipment. One example, meant to be 'Ortega', the name of Raudive's former teacher, just sounds like screeching feedback. Another, supposedly a message delivered in Spanish, could equally be interpreted as being song lyrics complete with flamenco music in the background.

Despite its promising title, then, *Breakthrough* doesn't do much to change a sceptical view of electronic voice phenomena, and really only provides further examples of how pareidolia can cloud our judgement when it comes to interpreting potentially uncanny stimuli. While the recordings are creepy in their own right, their nonsense phrases and, occasionally, musical tone suggest that Raudive was catching snippets of radio stations. On the other hand, Raudive acknowledged this in a way that, although ludicrous, is as difficult to disprove as it is to prove unless we have access to the radio programme being broadcast live in the moment Raudive was recording his samples: he suggested that ghosts have their own radio stations, and that when he picked up music or radio-style announcements, it was actually from a beyond-the-grave DJ.

Even though such technologies are commonplace today, we're still susceptible to being fooled by the 'evidence' captured through digital media. It remains a worldwide phenomenon; in Japan, for instance, photographs that include a ghostly 'extra' are an urban legend known as *shinrei shashin* ('ghost snapshots').[12] Rather than the professional portraits done in the nineteenth century, these are spontaneous photographs taken on smartphones and cameras, often made by tourists or at family gatherings and personal moments such as birthday parties. A *shinrei shashin* cannot be deliberately taken; the 'extra' often appears when they are least expected. When the photographs are viewed (or, back in the disposable camera days, developed), one may well feature an extra person who wasn't present when the shot was taken. Magazines and media have helped to create a cultural phenomenon from *shinrei shashin* by dedicating pages and segments for the general public to submit their photographs. However, where the spirit photography of the era of Spiritualism was revered and treasured by the sitters, *shinrei shashin* are more often destroyed than kept. The ghosts that choose to appear in these photographs are said to be malevolent, and it's better to dispose of the evidence.

It's likely that as forms of digital recording become more advanced and more common in the form of live-streaming and CCTV doorbells, so too will reports of 'evidence' of ghostly activity. But recordings can never really prove anything; all we're seeing or hearing are signals, pixels. Moreover, convincing deepfakes and AI-generated images mean that more than ever we're unable to trust what otherwise seems like an objective form of data.

Nevertheless, the spirit photography of the nineteenth century began a trend for capturing ghosts in a way that made the intangible somewhat more physical. It brought the dead that little bit closer, but, of course, it still wasn't enough for Spiritualists. They wanted not only to see the spirits of the dead back in the land of the living, but to look further into the place those spirits returned to after their brief visitations in séance rooms and photography studios.

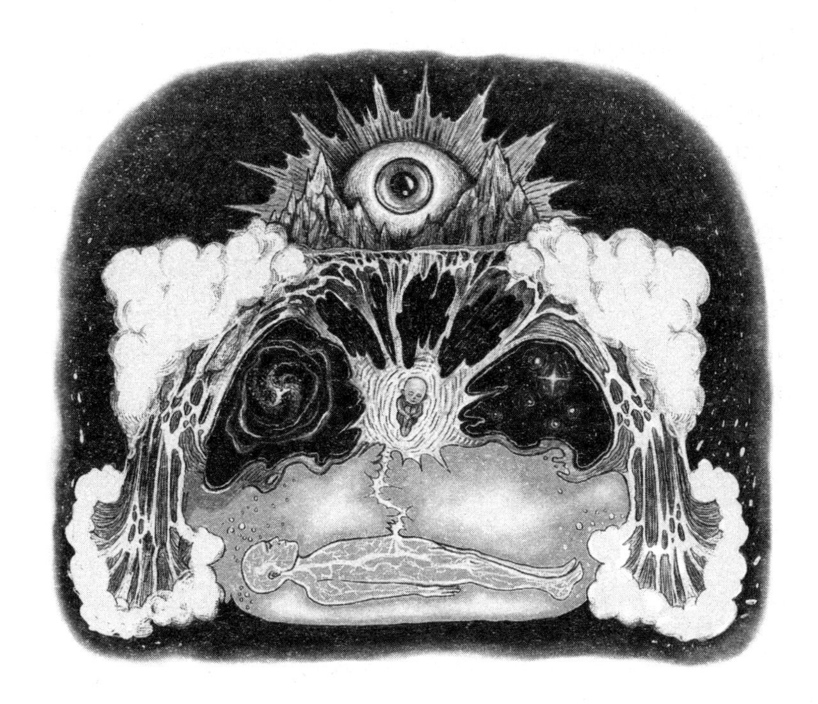

CHAPTER FIVE

Summerland

Spiritualism and science were closely intertwined in the nineteenth century. As we've already seen, the techniques mediums used and the ways in which Spiritualists described their evidence were intrinsically linked to scientific advancements such as radio waves and photography. But Spiritualism was, after all, a religion too, and was just as heavily influenced by Christianity and other world beliefs.

There was one scientific revolution that shook religion deeply during the height of Spiritualism: in 1859, barely a decade after the Fox sisters began to speak to the ghost of Mr Splitfoot, Charles Darwin published his study of evolutionary biology, *On the Origin of Species*. Using observation in context with the growing fossil record, Darwin famously posed the theory that all life on Earth has evolved into, and out of, its wonderful varieties over millions of years due to environmental pressures. Its biggest claim, many thought, was that it essentially disproved the Christian story of Creation. God could not have created the world and plonked Adam and Eve on it, because humans didn't come from nowhere: according to the theory of evolution, we came from an apelike ancestor which we share with other apes, including chimpanzees. Just like everything else on the planet, we evolved from other forms of life. Darwin's book had a profound effect on the Christian faith, the leaders of which were suddenly being asked to reconsider their place in the universe.

It had a surprisingly big effect on Spiritualists, too. They wondered about the nature of animals, and of souls, and, more importantly, about where evolution was heading. If humans weren't God's creation, and were merely one step in a greater timeline, then it made sense to the Spiritualists that the survival of the soul after death was the *next* step in each individual human's evolutionary timeline. We evolved into humans from primordial ooze, and in death we evolved again into something else.

This idea quickly took root in the séance room. Towards the end of the nineteenth century, the mood had shifted in relation to spirit manifestations. Raps and knocks were old hat by this point, and people wanted more than for spirits to declare their presence, wave a trumpet around their heads or scribble with a planchette. The public's curiosity, especially now that the idea of God was under more strain than ever, moved in a different direction. Spiritualists were satisfied that they had received their answer that the soul did persist after death; now they wanted to know where those souls went.

Spiritualists, then, began to create their own idea of heaven. It was a utopia that embodied their other concerns such as equality and socialism, but also demonstrated the scientific influence that continued to be felt so strongly among them. It was given many names, but the most popular was 'Summerland', a name with a slightly pagan flavour that suggested an afterlife as carefree as a warm August evening.

The term was coined by American spiritualist Andrew Jackson Davis, a man who had been wholeheartedly involved in Spiritualism from its earliest days in the Fox sisters' house, and who believed he could clairvoyantly see the realm to which we moved after death. He wrote extensively on his 'findings', most notably in two books, *A Stellar Key to the Summer Land* (1868) and *Views of Our Heavenly Home* (1878). Both texts set about explaining Summerland in inherently scientific terminology, clearly influenced by the way certain discoveries in evolution, but equally in astronomy, were challenging religious notions of the afterlife. By the 1860s, Spiritualism had become deeply embroiled in science; while some scientists were trying to refute the reality of séances and ghosts, others, such as the eminent French astronomer Camille Flammarion, openly advocated for the veracity of Spiritualism. Indeed, Flammarion's influence is clear in Jackson's writing, and both books read like obscure cosmological textbooks at times – it's easy to forget that Davis is talking about an ethereal afterlife rather than a newly discovered planet.

The way Davis saw it, Summerland exists above us, somewhere in the Milky Way, which, at that time, was seen as the full extent of the universe. Inspired by the cosmic dust that orbited Saturn in a ring, Davis envisioned Summerland as having a similar structure. He called the afterlife a 'stratified zone', a habitable and physical belt made of mysterious and ethereal matter.[1] Atoms, according to Davis, exist in a hierarchy – the more 'refined' the organism, the more special are the atoms that compose it. Humans are made of the 'most refined particles', and when our bodies disintegrate, these particles ascend to form the matter which composes the afterlife Zone. Davis, on one of his clairvoyant jollies to Summerland, marvelled at the rich vegetation, fertile soils, the flowers and forests that furnish the land – all made up of atoms that were once human bodies. And each day, he claimed, millions of tons of these atoms float out into space from Earth to collect in the Zone. In other words, Summerland is both inhabited by dead people and composed of dead people; an astral belt through the Milky Way made up entirely of once-human atoms. Every time you exfoliate, you plant a tree in Summerland. That's nice.

This hierarchical structure of atoms is reminiscent of Darwin's theory of evolution, which argues that organisms often become more sophisticated over a period of time. But Davis didn't leave it there. He argued that Summerland itself is hierarchical, that those 'refined' particles of human bodies, now forming pure spiritual matter, is also hierarchical. Through a scientific-looking diagram, Davis demonstrated the hierarchical planes of Summerland, from its most basic level on Earth to the seven 'Spheres' in which some sort of unnamed and unknown deity (but not, he seems to insist, the Christian idea of God) resides. Again, we see the constant blending of science and religion in Spiritualism that would only become increasingly complicated.

Having proven through astrophysical means, at least so he thought, the existence of the Summerland belt, ten years later Davis's follow-up book, *Views of Our Heavenly Home*, presented a more detailed picture of the afterlife through his clairvoyant visits. I thought *A Stellar Key to the Summerland* was one of the most bonkers books I've ever read, but this sequel is truly something else. Davis claims that, between books, he had further developed his ability to 'live in both worlds naturally and healthfully' through his clairvoyant sight that acts as a spiritual telescope.[2] He states that Summerland is identical in geology and architecture to Earth, but also draws a delightful map of Summerland's lakes that looks like something out of a high-fantasy novel, with names such as the Seven Lakes of Cylosimar.

Davis also describes a very peculiar scene that makes me wonder if his clairvoyance was, instead, some sort of dream-like state which he was convinced gave him virtual access to the afterlife. In Summerland's 'rainbow-tinted' version of New York, everyone seems to have some sort of animalistic feature (again, perhaps influenced by Darwin here) that represents a particular characteristic of their personality. One man has a horse's head, another the head of a lamb; he sees a handsome gentleman with the legs of a goat, while someone else has a dove constantly perched on their shoulder. Presumably, then, as we move through the rest of the planes of Summerland, we shed those animal parts of ourselves. He never came back to this theory, though, and soon returned to a more scientific view of the spiritual Zone.

He also turned his attention to the journey itself from Earth (which he now called the Winterland), to the astral plane of Summerland. His keen interest in astronomy and the Spiritualist-scientific thoughts of Camille Flammarion once more take centre stage here, and he explains how we pass from Earth to the belt of spiritual matter without dying a second time in the vacuum of space. In death, our spirits, once internal, are now external. What we felt in our crude Winterland bodies no longer affects us. Davis writes, 'But we do not go out after death with these chronothermal nerves; hydrogen is not our after-death envelopment; the spiritual body is impressible by nothing less fine than that omnipresent solar-influence and astral-ether which we have agreed to name *the Spirit of God!*' We don't feel the cold of space nor the burn of the Sun, nor, interestingly, 'the decomposition of countless universes in eternity'. The first black hole wouldn't be discovered for almost a hundred years after Davis's book, but no doubt he would have claimed that the Summerland belt is immune to such an immense gravitational pull, too.

And then things get even weirder. In the latter half of the nineteenth century, the Victorians developed a mild obsession with aliens. Apart from H. G. Wells' 1890s text *The War of the Worlds*, aliens and extra-terrestrial life aren't really something we tend to associate with Victorian culture, but there is a surprising abundance of writing on the subject from a Spiritualist point of view. Ideas of Summerland, a great cosmic afterlife, became embroiled with theories about aliens. Astronomy, while advancing quickly, still didn't have equipment powerful enough to understand whether other planets in our solar system were capable of housing life. Imaginative Victorians, rarely known to underestimate the nature of things, began to dream up sophisticated civilisations on other worlds. For the Spiritualists this was a fascinating concept, and one that blended delightfully with their ideas of a pseudo-scientific cosmic afterlife. Through his clairvoyant telescope, which was of course far more advanced than any of the most renowned astronomers' equipment, Davis took a look at the civilisations living on the famously hospitable planets Jupiter and Saturn. He called the reality of their existence an 'every-day observation', and described the inhabitants of both planets as being far superior to our own. Perhaps the scale and patterning of Jupiter and the unusual rings of Saturn made them both seem somehow like more advanced planets than Earth. Davis described these extra-terrestrial lifeforms as having 'exceeding refinement, purity, and interiority'; and, because of their higher intelligence and sensitivity compared to humans, argued their respective versions of Spiritualism were far more widespread and believed. Indeed, he believed both Saturn and Jupiter naturally operated with an acceptance of spiritual communication. While these civilisations resemble human society but in a more futuristic sense, Davis almost seemed to suggest that the spiritual strength of these planets, and of others in the universe, have more of an influence on the shaping of Summerland than Earth: 'parts of it resemble Saturn's scenery more than ours; while other sections, unspeakably more perfect, exceed in harmony and loveliness anything known or imagined upon this or any other planet in the universe.' In reality, Saturn doesn't have a solid surface and it's made up of very cold, highly pressurised gas. Not particularly harmonious.

Spiritualists continued with this strange idea of alien afterlives. As late as the 1930s, when the structure of the universe was better understood than in Davis's day, books describing Summerland further explored the celestial spheres in relation to astronomy and extra-terrestrial life. In *Spirit-World Teachings*, Shirley Carson Jenney wrote down messages from her deceased mother about life in Summerland in the form of questions and answers. Jenney asked her mother whether the Moon has its own afterlife sphere, to which her mother responded that the Moon 'has a ringed Heaven, and it is of three-divisional aspect. Its angels are most beautiful; they are here with us sometimes.'[3] Moreover, Venus was said to have 25 spiritual planes attached to it. Davis, however, argued that the Moon was without life, and therefore didn't contribute to the ever-growing celestial afterlife in the Milky Way like Jupiter and Saturn did (but predicted that, one day, we would inhabit the Moon and, inevitably, its own band of spiritual dust would form as a result).

But aliens weren't the only unexpected occupants of Summerland. In 1873, Andrew Jackson Davis was continuing to build upon the increasingly peculiar lore of the Spiritualist afterlife, and published a text called *The Diakka, and Their Earthly Victims*. His idea of Summerland was becoming something akin to the popular traveller's narrative, and even seems to parallel aspects of Lilliput from Jonathan Swift's *Gulliver's Travels* (1726) in relation to its fantastical landscapes and inhabitants. Somewhere near the constellation Draco, said Davis, exists a part of Summerland reserved solely for a strange race of people called the Diakka. Again, consisting of a belt of cosmic dust, the Diakka's country was calculated by Davis to be nearly two million times the diameter of the Earth. The Diakka were described by Davis as 'morally deficient and affectionally unclean'.[4] It wasn't Hell as such because he made it sound like they're all having a party up there – nor was it a form of purgatory. It was simply a separate territory for those who rejected the peace and gentle, communal spirit of the rest of Summerland. Diakka were once human spirits from across the world who, after death, found themselves changed into something powerful and mischievous because of their Earthly habits and beliefs.

As he continued to describe the Diakka, Davis's motivation for expanding his mythology became apparent. Crucially, by 1873, Spiritualism had already been tarnished by scientists and critics exposing fraudulent mediums. It was also the year that Florence Cook's materialised spirit Katie King was seized by Mr Volckman and her authenticity was cast into doubt. The Diakka became a scapegoat to explain away these blights upon Spiritualism in a manner that still legitimised the practice and, therefore, the existence of Summerland. These spirits were stronger and far more able to bridge the gap between Earth and the cosmos, and often did so in order to cause havoc – especially in séances and sittings with mediums. They affected everyone on Earth with their malevolence, but seemed particularly hell-bent on undermining Spiritualist activity for unclear reasons. Indeed, Davis gave the case of Florence Cook and Katie King as his prime example of what he felt to be the Diakka in action. Florence Cook was being influenced by a Diakka into impersonating a spirit; it wasn't the fault of a Spiritualist medium for being exposed as a fraud – it was simply a Diakka using their susceptibility to influence them into practising fraud. In a way, Davis was predicting the modern rise of demonic forces in amateur paranormal investigations. We'll examine these in detail in a later chapter, but the vocabulary of today's paranormal investigators frequently features notions of dark and demonic entities who cause confusion, chaos and malfunctioning equipment. It shows, I suppose, the way Spiritualist lore continues to evolve to explain the unexpected and unforeseen.

It became a preoccupation among Spiritualists to investigate and find out as much as possible about this wonderful new type of afterlife. Séances were held not simply to establish contact with the dead, but to have lengthy conversations in which Summerland was described in as much detail as possible. The more detail they received, the more likely they could get a book out of it. Indeed, many books were published at the turn of the century that purely consisted of describing the Spiritualist afterlife with, as one might expect, often contrasting ideas and details. No two books about Summerland are the same, and no two books describe Summerland in exactly the same way. The stark differences between what spirits experienced after death rather raised more questions and scepticism than the Spiritualists believed they answered.

One of the most interesting examples of these books was 'written' by the spirit of the well-known Irish journalist and influential newspaper editor W. T. Stead. When Stead departed forever for Summerland in 1912, he was on his way to a peace congress in Carnegie Hall, New York, on the doomed RMS *Titanic*. In the late hours of 14 April, the ship struck an iceberg and sank a few hours later. Having given his life jacket to another passenger, Stead was last seen by a survivor clinging to a life raft before the bitter cold of the Atlantic Ocean set into his limbs and compelled him to let go. Out of around 2,200 passengers, Stead was among the 1,496 who died.

In life, Stead was a fierce advocate of Spiritualism and the notion of a pseudo-scientific afterlife. As a member of the Ghost Club, he often wrote about ghosts for the newspapers he worked for, including *Pall Mall Gazette* and *Review of Reviews*, but eventually set up his own (albeit short-lived) Spiritualist publication in 1890 in order to write without the constraints of sceptical newspaper editors. It was called *Borderland* – a title that harks back to the notion of a utopia beyond death.

It wasn't simply his articles about Spiritualism that made him well known in occult circles; Stead seemed to be eerily aware of the violent fate that awaited him. In 1886, he wrote a short story called 'How the Mail Steamer Went Down in Mid Atlantic, by a Survivor'. After his death, the work exalted him in the eyes of his Spiritualist allies, suggesting that Stead had predicted his own demise. It wasn't the only example, though, although it remains the most notorious. In 1909, just three years before his icy death, Stead republished in book form a long article he had written for *Review of Reviews*. Titled 'How I Know the Dead Return', it is an impassioned manifesto that explains his certainty of the existence of Summerland. While his short story about the mail steamer that seemed to foresee the *Titanic* disaster remains the greatest legend about Stead's death, this article shouldn't be ignored for its uncanny predictions. In describing the journey to Summerland, he uses an obscure metaphor that is quite chilling: 'Let us imagine the grave as if it were the Atlantic Ocean.'[5] Within three years, it would be. He seemed to be absolutely convinced that he would meet his end through drowning, so clearly it was natural to see the afterlife as a vast ocean on which he was forever floating.

Stead entreated the scientific community to investigate the survival of the human soul after death, basing his conviction on the fact that Summerland hadn't yet been absolutely *disproved*. 'Put the percentage of probability as low as you like,' he wrote, 'if there be even the smallest chance of its truth it is surely an obvious corollary from such an admission that there is no subject worthy of careful and scientific examination.' Indeed, he even went so far as to call it 'unscientific' that scientists weren't putting effort and resources into investigating even an infinitesimally small chance of an afterlife. This is something I keep thinking about as I write this book: there are so many millions of anecdotes and cases of hauntings around the world, spanning millennia – all it would take is for just one of them to be a genuine incident of a person's spirit existing beyond death, and it would surely change everything we know about the universe. Isn't that worth investigating?

Stead's certainty in Summerland came not from his own foreboding of death, however, but, as we've seen time and time again, from grief. A year prior to writing his article, his eldest son, Willie – the son whom he had nurtured into following in his footsteps – died. 'The tie between us was of the closest,' Stead wrote, and the pain of loss, even in his belief of an afterlife, is palpable in the article. Almost immediately following his son's death, he began to receive messages from Willie in séances. While there's not much detail regarding what was said, Stead impressed upon his reader the fact that he was absolutely certain of his son's personality and style of writing, and that it would be impossible for a medium to know half the things that were shared by Willie in the messages. He ended his article with a passionate conclusion that would lay the foundations for the aftermath of his own death:

> After this I can doubt no more. For me the problem is solved, the truth is established, and I am glad to have this opportunity of testifying publicly to all the world that, so far as I am concerned, doubt on this subject is henceforth impossible.

Stead's involvement with Spiritualism and his frequent attendance at sittings had allowed his children to grow up in an environment where such activities were commonplace and encouraged.

His daughter, Estelle, had been on tour with a Shakespearean theatre troupe while her father was aboard the *Titanic*. Among the members of the Company was a man named Mr Pardoe Woodman, who, hours before the ship sank, ominously predicted the disaster and told Estelle that an older gentleman connected to her would suffer fatal consequences.

It seems as though Estelle, even if not actively engaged in Spiritualism, was certainly open to its ideas and practices. Upon receiving the news of the sinking, and of her father's presumed death, she struck up a close friendship with Mr Woodman, whom she believed had a connection with her father's spirit.

Following his death, Estelle's need to reach her father beyond his watery grave grew with a sad intensity, and she began to actively seek out more information through mediums just as Stead had chased after the ghost of her brother Willie. A fortnight after the *Titanic* sank, Estelle visited a medium named Mrs Etta Wriedt, renowned for using a technique known as 'direct voice'. These mediums would fall into a trance, allegedly allowing the spirit to almost possess their bodies and use their vocal cords to speak. They were adept at changing their voices, putting on accents and adopting mannerisms to make it seem as though it was, in fact, the visiting spirit talking through them. Estelle claims that she spoke for 'over 20 minutes' with her father through Wriedt's powers, and that it was the tipping point for her own investigation into what became of her father after his death.

The disastrous sinking of the *Titanic* was soon overshadowed by the eruption of the First World War, and Estelle's fellow troupe member, Mr Woodman, was enlisted into the army. By 1917, he had been badly injured and was permanently discharged through invalidity. As we will see in the next chapter, the war had a profound effect on the general public, who turned to Spiritualism like never before. For Mr Woodman, though, his preoccupation remained fixed on the spirit of W. T. Stead, who seemed to have been haunting him since the *Titanic* sank. Now back in England, and struggling to adjust both to his experiences on the battlefield and his life-changing injury, Mr Woodman started to explore his mediumistic powers. He began to practise another technique of mediums: automatic writing. Like the direct voice, this allows the spirit to take charge of a medium's body to communicate, but rather than speak through the medium's mouth, a spirit controls a pen in the medium's hand to dictate as many messages as possible.

For Woodman, it was W. T. Stead, the great journalist, who still had much more to write. He and Estelle began to have sittings together, in which Woodman, with his eyes closed or covered with a handkerchief, would scribble furiously as Stead began to reveal his experiences after the sinking of the *Titanic.* It gave Estelle immense hope, who felt that 'the link between us is even stronger to-day than in 1912, when he threw off his physical body and passed on to the spirit world.'[6] Clearly, it was the spirit world that fascinated her most. Her final memory of her father was seeing him wave to her from the deck of the doomed ocean liner, but knowing now what sort of fate had befallen him, her imagination must have conjured up some awful scenes. Even I shudder to think of the way the water must have numbed his grip on the raft, the slow realisation creeping in like the cold that he would soon slip down into the depths of the Atlantic. And there his body remains; there couldn't be a proper funeral or burial – Estelle had to live with the knowledge that her father's final resting place was at the bottom of the sea, surrounded by other icy-blue corpses and the wreckage of a once-splendid ship. It's not difficult to understand that she would cling, just as her father had done to the raft, to an idea that her father was *not* at the bottom of the Atlantic, that his final moments were *not* of unimaginable cold, of water seeping into his lungs and of the chaos and screams of other suffering passengers – but that he had moved on to a beautiful, peaceful place and was there waiting for the rest of his family.

Her father's ghost, through Woodman's pen, decided he wanted to write a book. As a way to comfort Estelle and similarly bereaved people, but also as a way to continue his career as an investigative journalist, 'Stead' embarked on a new journey through this mysterious world of the dead. The book was called *The Blue Island: Experiences of a New Arrival Beyond the Veil* and was published in 1922, described as having been 'communicated' by W. T. Stead and 'recorded' by Pardoe Woodman and Estelle.

Stead began with the moment of death itself, but did not seem to want to linger: his focus was on his investigation into his new home. Through Woodman, he wrote: 'Of my actual passing from earth to spirit life I do not wish to write more than a few lines. I have already spoken of it several times and in several places. The first part of it was naturally an extremely discordant one, but from the time my physical life was ended there was no longer that sense of struggling with overwhelming odds; but I do not wish to speak of that.' Nevertheless, he was one of around 1,500 to lose their lives in the disaster, all floating up at the same moment to the mysterious Summerland. Stead told Mr Woodman and Estelle that his first instinct was not of fear or grief for his own death, but one of an overwhelming 'position to help others' as they left their bodies alongside him. But he only knew he was dead because, suddenly, he was surrounded by friends who had already died over a number of years. 'Oh,' he wrote, 'how badly I needed a telephone at that moment! I felt I could give the papers some headlines for that evening.'

Ironically, and perhaps a little morbidly, he described the way the *Titanic*'s dead waiting for the last soul to depart was like 'waiting for a liner to sail; we waited until we were all aboard.' The 'strange crew' were in varying states of emotional turmoil, but Stead seemed strangely calm around them. Eventually, they left, looking down on their own blue bodies in the water. Recalling both a Christian tradition of heaven and Davis's notion of the band of Summerland in the Milky Way, the spirits all departed upwards, towards the sky.

They seemed to be journeying for a long time, and then, finally, they saw a new land in the distance. Once they reached the land, each person was greeted by an old, deceased friend or relative, who helped them take their first step in Summerland. For Stead, it was his father, 'dressed as I had always known him.' I'm surprised at this; having read Stead's impassioned description of the séances with his son, Willie, in his article for the *Review of Reviews*, and the clairvoyant connection he believed to have had with him, surely it would have made more sense for Willie to have been the one to welcome him to the afterlife. Even more perplexing was that as soon as Stead was greeted by his father, they went for 'refreshment at once.' He didn't say what this refreshment involved, but it's wild to think that you get a cup of tea on arrival at the afterlife. Stead's father then began to walk his son through this strange place, which Stead described as having a strange blue aura to it. 'I do not mean,' he wrote through Woodman, 'the people, trees, houses, etc., etc., were all BLUE; but the general impression was that of a blue land.'

It is here that the description departs from its quasi-religious influence, and moves into more interesting territory. As is common with books about Summerland 'written' by its residents, Stead described a utopian community in which all are equal in toil and pleasure – where everyone works on a common task, but is also free to engage in personal development and interests. But the most important thing on arrival is that the newcomer processes the shock and grief, and cuts their ties with life back on Earth. His description of the afterlife handily paralleled his prediction of death in his article, 'How I Know the Dead Return', in which he used the metaphor of the Atlantic Ocean. Each departed spirit navigates this ocean, from one island to another. Being both Irish and a writer, Stead was likely aware of the traditional Immram narrative structure – old Irish stories in which a pilgrim went on a journey of religious discovery or redemption by travelling by boat across a series of otherworldly islands. Every island in an Immram story posed its own challenge by being inhabited by a demon or a saint, ready to test the protagonist. This is how Summerland is described in *The Blue Island*; the first port of call is almost like a luxurious holiday, a place of respite and leisure. 'There are houses', Stead told Woodman and Estelle, 'given over to book study, music, to athleticism of all kinds.' It is a place to exercise the mind and body, while allowing the newcomer to adjust to their new life.

The Blue Island is not the entirety of Summerland, but a kind of anteroom before spirits move once more to a permanent residence – just as Davis described multiple spheres of ascension through the afterlife. Once the spirit has acclimatised, they move to what Stead called the 'Real World'. In a similar way to how Stead first arrived at the Blue Island, spirits move upwards in one big group at a time, as though they're cramming into a lift. The Real World, Stead's Summerland, is so similar to Earth that for a moment he was disappointed; he mentioned that everyone was bustling about their routines as normal, and that the landscape doesn't have the blue lustre of the first plane. But life is different for everyone; they all make their own heaven in relation to their own tastes and their own perception of comfort. Houses are situated and shaped dependent on personal preference – he described how a person who lived in poverty on Earth now lives in an elaborate palace.

Again, however, there is the notion of work – and even a concept of currency. The life spirits lead in the Real World is luxurious, but they have to work to maintain that luxury. Stead spoke of 'progress'; each must work on their own spiritual and intellectual development through cultural pursuits and scientific research (for which there is a dedicated 'House' in the Real World), in order to earn the objects and lifestyle they desire. All beloved pets and animals lost over the years also make their way to the Real World, but presumably cats, as in life, are allowed to experience the height of luxury while doing absolutely nothing at all to earn it.

Most books about Summerland 'recorded' from messages received through séances end with moral encouragement for those still living on Earth, and *The Blue Island* is no different. Stead's message to Estelle, and to his readers, was that Earth is 'the training school' for permanent life in the Real World. We must prepare ourselves for that eternity of spiritual development through understanding our characters, our likes and dislikes, our hobbies and passions, so that we're better able to pick up where we left off when we shed our mortal bodies.

The description of Summerland allegedly given by Stead's ghost does not much engage with the debate that was ongoing in both the scientific and Spiritualistic communities: what evolution tells us about our place in the universe. Two years before Pardoe Woodman and Estelle Stead published *The Blue Island*, another investigation into Summerland aimed to answer some of these philosophical questions. J. S. M. Ward's *A Subaltern in Spirit Land* (1920) seems to have been written to unite Darwin's theory of evolution with the Spiritualists' notion of the afterlife.

Again, however, it was borne out of grief. Ward explained how his investigation began with the death of his brother in the First World War. As Estelle was pained by imagining her father lying at the bottom of the Atlantic, so too did Ward struggle to accept that his brother met such a terrible end. He was desperate to believe there was something more. Ward didn't approach mediums, however; he found that he had mediumistic talents of his own. He began to have what we would now call out-of-body experiences; in sleep, his felt as though his spirit departed from his body to join with his brother, Rex, in Summerland. He seemed to find that he could come and go freely, but always with a secure connection to his material, living body still on Earth. He was like a visitor to a hospital or care home. Almost every night, he went to find Rex as he adjusted to his new afterlife, and learned about what he had been doing and how he had been getting on. There was no immediate peace for Rex in Summerland. Ward found that Rex was depressed by what happened to him; he told his brother that he was 'fearfully lonely' in Summerland, despite having the unique experience of a brother who could astrally project himself into the afterlife.

Apparently, according to Ward's description of Summerland, when we go there we largely become accustomed to it quite well. Rex, however, had unusual difficulty accepting his new existence. This is where things become particularly odd, and show just how much of a hold the theory of evolution had on the public imagination, especially when it came to reconfiguring ideas of death and the soul. Because Rex was being a problem patient, the brothers' deceased uncle, H., appeared like Gandalf and said that Rex needed, essentially, a spiritual time-out. It all sounds a bit sinister; Uncle H. told the brothers that Rex was being made to go to the 'Land of Utter Silence', so that 'his troubled spirit may recover its equilibrium.'[7] This functions a little like Stead's Blue Island, an anteplane in which the newcomer can process their own death and adjust to the prospect of eternal afterlife. As we've seen already, while these books about Summerland share some similarities, their differences are stark and contradictory. No spirit describes Summerland in exactly the same way – but, of course, these discrepancies did nothing to sway Spiritualists' belief in this higher plane of existence.

The three set off together, and the book becomes a kind of *Divine Comedy* for the post-Darwin public. In the fourteenth-century text by Dante Alighieri, Virgil comes to take the poet on a tour through Hell, Purgatory and Heaven; Dante sees all sorts of strange and horrible sights, from a minotaur to a frozen pool full of ne'er-do-wells gnawing on each other's heads. At the end, Dante emerges back into the night, full of renewed appreciation for God, his own mortality and the deeds he has yet to accomplish. *A Subaltern in Spirit Land* works on much the same principle, but instead of moving through the various planes of Christian existence, they move through a bizarre history of the universe as influenced by Darwin and contemporary astronomical discoveries of spiral galaxies.

Their uncle, H., told Ward that they must pass through the astral plane, where all the Earth's stages continue to exist like separate 'rooms' of time. He planned to walk them back through the Earth's history to the 'Archaean Rocks' – a silent, barren place 'practically devoid of all life' where Rex could rest and recover from his traumatic death. They started their journey, and were soon accosted by the spirit of a man in a Roman toga, who asked in a very authentic Roman sort of way: 'Whence came you, strangers, and whither going?' Ward seemed bemused by the historical figures they came across, but soon they left modern humans behind and moved into the Neolithic and Palaeolithic Age where some Neanderthals gave them a bit of bother.

Then comes my favourite bit. After they left the Palaeolithic Age, they returned to the time of the dinosaurs. Yes, that's right: all dinosaurs *did* go to heaven. They get attacked by some ghostly pterodactyls, naturally. Once back from their adventure delivering Rex to the Land of Utter Silence, Ward and his uncle spent some time exploring the main part of Summerland, where most spirits roam. The picture is of a modern, middle-class utopia. Ward's uncle explained that there were hundreds of shops that carried the 'ghosts' of broken items; whenever a 'curio' was destroyed in the material world, its spirit form appeared in a shop. There were magnificent gardens, art galleries and, Ward wrote, 'you'll be delighted to learn there's a very good museum' (I am, but I have a very important question: is there a palaeontology section, and do the dinosaur ghosts visit their own skeletons?). There was, briefly, a zoo in Summerland, but the spirits of animals easily phased through the bars of their cages and all that was left were three lazy monkeys. The lion was one of the first to escape, who 'created quite a panic walking down the High Street of our peaceful little town.' Not to pick holes in this otherwise sound and logical explanation of the afterlife, but why would everyone panic over a loose lion if they're already dead?

Animals are often mentioned in books about Summerland. Unlike in Ward's text, however, where we see the past evolution of life still living in the various spheres, some books posed the idea of evolution continuing to play out in the heavens. Earlier in the chapter, we looked at Shirley Carson Jenney's *Spirit-World Teachings*, in which she spoke to her late mother in a series of sittings. Her mother seemed to be having some sort of frisson in Summerland with the ghost of Percy Bysshe Shelley (doubtful not least because the poems her mother sent via automatic writing were extremely terrible), and who described the usual utopian wonderland. However, during one séance she described a number of flamboyant and exotic-looking birds that existed only in Summerland, their spirits evolving into higher forms on the plane. There was the yellow-and-black songbird called a Marshmay, a little scarlet bird called a Thumbkin, the nesting Grellon-bird, and a 'wee green bird with red eyes, and a beak very curiously coloured' called a Pippo. But if these birds were so remarkable and delightful and vivid, why haven't we heard of them before in other books about the afterlife? Why doesn't Andrew Davis Jackson, with his clear view of Summerland from his spiritual telescope, tell us about these birds? Surely Stead, with his journalist's flair for description, would have delighted in mentioning them to his daughter?

This is the pitfall of so many of these books about Summerland. Each tries to be more detailed, more unique, than the last – the selling point is the fact that they are sharing more information about the afterlife and that their medium-author is stronger and more knowledgeable than any of the competitive books that sought to do the same thing. Because of this, there's always a point at which they pass from eerie to downright ludicrous. For Ward, his description of concepts such as the museum and zoo provoke more questions than they answer. In Margaret Cameron's *The Seven Purposes: An Experience in Psychic Phenomena* (1918), it's the notion of 'disintegrating forces' that are confusing and difficult to grasp. Once again employing scientific terminology, her conversations with relatives describing their life in Summerland returned again and again to a form of spiritual physics. Disintegrating forces were, essentially, bad vibes that damage 'spiritual progress'.[8] These forces most commonly originated on Earth, and moved up to contaminate Summerland. The work of the spirits – again, we see this utopian concept of equal and earnest toil among spirit-folk – was to counteract the disintegrating forces through sending messages of love and hope to those still alive. The more people who believed in Spiritualism, the more those nasty disintegrating forces would be quelled.

Despite the contradictions in the idea of Summerland, it was still widely believed and accepted by staunch Spiritualists. The religion rested, of course, on the notion that there was an afterlife, but the developing picture of a cosmic utopia somewhere in the Milky Way, in which spirits evolved as life evolves on Earth, gained particular traction among some of its most fervent believers. This was especially the case in the US, where Spiritualism was causing new branches of theology to form and create new beliefs that were Christian in origin but pseudo-scientific in scope, with questionable doctrines and strict hierarchical structures disguised in notions of equality. In other words, cults.

In the last decade of the nineteenth century, Andrew Jackson Davis's notion of cosmic spheres and an afterlife that existed quite literally among the stars garnered quite a following in the US. California seemed to be an especial hub for Spiritualism, with its mediums and circles focusing particularly on the afterlife. The San Francisco-based Spiritualist journal, ingeniously titled *Golden Gate*, presented itself as a 'journal of practical reform, devoted to the elevation of humanity in this life, and a search for the evidences of life beyond.'[9] As well as frequently promoting the work and services of Andrew Jackson Davis, who was still a practising medium when he wasn't gazing through his spiritual telescope, *Golden Gate* excitedly covered and advertised a developing commune on the outskirts of Santa Barbara called, of course, Summerland. It was billed as a 'Spiritualist Colony' that was much desired and anticipated among Californian circles and beyond, and was a place where Spiritualists could congregate, live permanently, and concentrate their powers and education in mediumship. The land itself, an ex-ranch, was described in much the same way as the spirits painted glorious, rich images of the afterlife, with 'a most beautiful view of the mountains, islands, ocean', rich and fertile soil, and 'pure spring water' from an 'unfailing source'. A quadruple plot required for building a residence was $120, or just over $4,000 (£3,139) in today's money.

Its objective was to 'advance the cause of Spiritualism', and it was described as being governed 'the same as other towns and cities', including a prohibition against alcohol. Already by 1889, when Summerland, CA was in its infancy, lots were being snapped up. Some Spiritualists had by then built their homes and moved in, and a wealthy benefactor had bought a plot and the materials with which to build Summerland its own library. Remarkably, Summerland wasn't a fleeting venture. In April 1932, the popular magazine *Spiritualist Monthly* called for its readers to join in the celebrations of Summerland's 40th anniversary.[10] At this point in the US, dozens of quasi-religions and cults had been set up, including the Spiritualist Success Church and the deeply problematic Christian Science churches. Nevertheless, good old-fashioned Spiritualism was alive and well in Summerland. Over the decades, the colony had progressed and attracted many notable international mediums, and the article described it as having 'a sense of invisible presences ever close – an uncanny something that inspires reverence in even the most sceptical.' The celebration sought to unite members of all the local branch sects that mingled Christianity and Spiritualism, and the colony continued to thrive until these sects fizzled out later in the century. Astonishingly, however, Summerland still exists as a popular holiday resort with a few thousand permanent residents. Santa Barbara's tourism website calls it a 'seaside haven' perfect for surfing and, much to the original developers' spiritual horror as they look down from the astral Summerland, a winery.[11] There's no mention of its origins as a colony for Spiritualists, although the website does recommend a visit to the Oprah Winfrey-approved Sacred Space, a shop and garden for the 'spiritually inclined', in a subtle nod to its history. It also promotes Summerland's 'good vibrations', although I presume this is a reference to the Beach Boys rather than Margaret Cameron's theory of 'disintegrating' and constructive forces.

By the mid-twentieth century, theories of Summerland had dwindled along with the rest of Spiritualism, replaced instead by practices of psychical research and parapsychology. Growing secularism in the countries where Spiritualism had its greatest hold likely contributed to this dwindling interest in exploring an afterlife, but, as we've seen, Summerland was too messy a concept to continue to seriously develop. Nevertheless, I think the motivation behind Summerland remains with us today. We're still trying to create our own private ideas of an afterlife, even if it's not with the structured enthusiasm of the previous century. Modern eulogies and obituaries are often phrased with what the bereaved hope the dead are surrounded by, even when it comes from the most secular position ('I hope there's plenty of beer where you are'). Headstones read that family members are 'reunited', which assumes the existence of a place where that reunion can happen. We even describe our pets crossing 'the rainbow bridge' to an afterlife, which is reminiscent of a poem written in 1959 by Edna Clyne who describes the utopia as having 'meadows and hills' and abundant sunshine.[12] While 'Summerland' is no longer a part of our mourning vocabulary, I think the concept very much still is. After all, death is just as painful, just as grisly and unfair, as it ever was, and I think the secular concept of a utopian afterlife is an important comfort to many – it means that a person's final moments are not of a crowded hospital ward and beeping equipment, that they still exist somewhere and continue to indulge in the habits that made them so memorable to their families.

Summerland, then, was a concept that originated from Spiritualists' desire to separate their ideology from purely Christian traditions. It combined the notion of heaven with contemporary debates about the cosmos, matter, forces and biological evolution. Where Darwin rocked the foundations of creationism, Spiritualists employed these new scientific theories to help them describe the survival of the spirit after death. While descriptions of Summerland began as a way for Spiritualists to stand firm against Darwin's new understanding of life and death, they soon took on a much wider cultural significance. Two years after W. T. Stead drowned in the Atlantic, the First World War erupted in Europe and led to the violent deaths of nine million military personnel – mostly young men in their first years of adulthood. The public mourned on both a personal and international scale, and many – some of whom had had no prior interest in Spiritualism – clung to the utopian ideal of Summerland as a way of consoling themselves that their sons were living peacefully after the horrors they had endured on the battlefield, and were waiting for their parents to join them in due course.

CHAPTER SIX

The Ghosts of War

The First World War began in July 1914, and by its conclusion in November 1918, over nine million soldiers had been killed in the conflict. Each death was devastating to relatives and friends left behind, and the first decades of the twentieth century were marred by individual grief on a global scale. The Victorians had turned death into something of a pageant, something to be celebrated as much as revered; for the large part, death followed illness and was often witnessed by surviving family members. The Great War, however, turned death into something sudden, cruel and unknowable, as young men – often barely into adulthood – were slaughtered on the front line, their deaths reported to their families days or weeks later.

In the first years of the twentieth century, Spiritualism had lost much of the popularity it enjoyed in previous decades. Admissions of fraud by the Fox sisters, numerous exposures of trickery and growing secularism meant that interest in communicating with ghosts was rapidly dwindling. The Great War, however, breathed new life into Spiritualism; as men began to die in their thousands, far away from their homes and families, mediums and investigators were afforded a reinvigorated sense of purpose. Séances ceased being a party game or stage act and instead became a beacon of hope for those who were left behind that their sons, brothers and husbands weren't truly gone. Spiritualism spread globally once more, with a new league of well-known and passionate supporters. Equally, though, it renewed the anger of sceptics; mediums weren't simply practising fraud now, but were exploiting public grief on a massive scale. The conflict in Europe was mirrored by a ferocious debate between scientists and Spiritualists, each side fuelled by the pain of their own private losses.

Among the nine million soldiers killed was Raymond Lodge, the youngest son of Sir Oliver Lodge. Lodge was a prominent member of Spiritualist circles and the Society for Psychical Research, as well as a distinguished scientist whose work formed some of the key early developments of radio communication. Raymond volunteered to join the conflict in France in September 1914, and by the spring of 1915 was in the trenches at Ypres. An ambitious young man who clearly inherited his father's quick intellect, he advanced through the ranks, and, taking over from an injured captain, was soon leading a company through No Man's Land.

On 14 September 1915, he was struck by a fragment of shell. His death took several hours, and his family received the news three days later.

Raymond was 26 years old.

Sir Oliver Lodge had been actively engaged in ghost hunting and attempts to contact the dead both before and throughout the war, and the rest of the Lodge family shared his interest. They were known, therefore, among the circles of the Spiritualist mediums whom they regularly visited to conduct séances and experiments. Barely two weeks after Raymond's death, the Lodges believed that he began to appear at their supernatural sittings. In 1916, Sir Oliver compiled his notes from these séances into a book titled *Raymond, or Life and Death.*

It is a strange, sad book. Usually, sources like this fascinate me; I scribble down notes with excitement at what I've come across, while my mind sparks with connections to other things I've read and how I might structure it in a particular chapter. My interest is mostly practical: how can I *use* the text? I type up the notes, return the book to the library or close the browser tab, and move on. *Raymond* was like nothing I've read before. It is an incredibly complicated and sensitive book, so much so that I found myself thinking about the Lodges, and Raymond especially, on my walk up to campus and while making dinner. I still think of Raymond when I pass the war memorial in the centre of my village. It's not really a book about objective experiments with various mediums, as much as Lodge presents it that way. It is a book about a father's crushing grief, and his desperation to follow any lead, any glimmer of evidence, that his son still exists somewhere – and that his final moments of consciousness were not of a slow bleeding-out in a filthy trench hundreds of miles from home.

Paranormal researcher Dr Kate Cherrell echoed my complicated thoughts when I spoke to her about Spiritualism during the First World War. It was, she told me, a period in Spiritualist history 'led by immediate and active grief'. She sees this era as something that helped to shape modern grief and attitudes to the afterlife. Whereas Victorian Spiritualists were 'propelled by zeal' and frequently mocked for it, Cherrell described how Spiritualist texts written from experiences of wartime grief are much more 'raw and immediate', much more about personal loss, and so provoke sympathy from even the most sceptical of readers.

Raymond is split into three parts. The final part is a rather rambling segment philosophising on proof of the afterlife – in line with dozens of similar texts from this era of psychical research. It is the first two sections that are the most interesting, and the most upsetting.

Lodge opens by collecting Raymond's letters together. He said he does this mostly for his family and for those who knew his son, and hoped that anyone else will forgive him for including them. In those letters, Raymond is truly alive. He is humorous, optimistic and light-hearted – even if it is only for his parents' benefit. 'I got a splendid reception from my friends here,' he wrote, upon arriving at the Front in late March 1915.[1] 'I am having quite a nice time in the trenches,' read a letter from 3 April. He described a cuckoo he often hears, and how much more notice he took of the night sky while keeping watch. Things began to affect him, though, despite the hope he pinned to the end of every letter. His friend Fletcher left for a rest cure – 'his nerves are all wrong' – and Raymond was upset without him. The roads became thick with mud, the numbers in his Company were diminishing, and on a night in June the Company advanced to dig a further trench, only to retreat again immediately having not made a lick of progress – but not without suffering the fatality of his friend, Thomas.

'Isn't it fairly sickening?' he wrote.

Raymond's final letter was dated two days before his death, describing an 'ordinary tour of duty' in the front-line trenches. He was excited to be going in a motor-bus.

The next correspondence the Lodges received was a telegram from the War Office. 'Deeply regret to inform you', it read, 'that Second Lieut. R. Lodge, Second South Lancs, was wounded 14 Sept. and has since died. Lord Kitchener expresses his sympathy.'

Lodge explained that Raymond's death and his involvement in paranormal research did not initially coincide. In fact, he wasn't even present when the first supposed appearance of Raymond occurred. A family friend, a 'French widow lady, who had been kind to our daughters during winters in Paris', had also suffered a recent bereavement in the war – both of her sons were killed, a week apart. Lodge's wife, perhaps trying to soothe her own grief through helping her friend, took her to sit with a medium. The séance took place on 25 September, when the loss of Raymond must still have been an open wound, and involved the techniques of table-tilting and rapping out letters of the alphabet. Amid messages directed to the French woman, something else seemed to come through:

TELL FATHER I HAVE MET SOME FRIENDS OF HIS

A name was given: Myers. Frederick William Henry Myers, one of the founding members of the most infamous ghost-hunting club of all: the Society for Psychical Research. Myers was particularly interested in ghosts and a potential afterlife, and worked tirelessly on experiments and controlled séances until his death in 1901. Lodge and Myers were close, and this, along with Lodge's interest in Spiritualism, was common knowledge. But perhaps it was the fact that Myers was now looming over the medium's table that Lodge threw himself into what he felt was an objective inquiry as to whether Raymond had genuinely spoken to him in the afterlife. It was what Myers would have wanted.

Lodge began to have séances with the medium Mrs Gladys Leonard, who was one of the most influential voices of the interwar Spiritualist movement. Famously, she held regular séances with the high-society lesbian couple Radclyffe Hall and Una Troubridge, who asked her to contact Hall's dead lover on a daily basis. Despite Leonard's notoriety, her substantial connections in Spiritualist circles, and the fact that it was announced in several major newspapers, we're promised by Lodge that she had no idea about Raymond's death. Over the course of several 'sittings', Mrs Leonard channelled

Raymond. His presence was fleeting and simple at first: his name was spelled out on the table. Soon, though, Raymond was chatting away through Mrs Leonard, and Lodge warned her that she might be overexerting herself.

'She pleaded', he wrote, 'that there were so many people who want help now [...] that she felt bound to help those who are distressed by the war.'

And who, presumably, paid rather well. A fraudulent medium of the time who was tried in the West London court belonged to the newly established College of Psychics, and told the jury she was paid an eye-watering sum of £50 a month, which in today's money is around £6,500. I think even I could channel the voices of the dead for that kind of salary.

The impact of *Raymond* was far-reaching. Within the first few years of publication, it ran to numerous editions. While it allegedly served to comfort soldiers on the Front, as well as their families back home, it also inspired an entire genre of books that evolved from the trend of Summerland texts, now written from the point of view of the ghosts of soldiers. Raymond's descriptions of the afterlife become increasingly detailed (and increasingly bizarre, claiming at one point that ghosts enjoy smoking ghostly cigars), as expressed through the automatic writing of Mrs Leonard. The technique was particularly popular during this period in the history of ghost hunting because of contemporary developments in wireless communications. Suddenly, Spiritualist texts used the new technology of radio signals to explain the role of the medium: they were the receiving device 'tuned in' to Spooky FM and able to hear, decode and record messages. Stewart Edward White, in his 1946 reflection on the wave of Spiritualism earlier in the century, *The Stars are Still There*, described this communication in terms of frequencies. Ordinary human beings in the 'Obstructed' plane of existence, he said, can reach a frequency of 100. Ghosts in the 'Unobstructed' universe, however, can only bring themselves down to a frequency of 150.

'The ordinary person – you or I – cannot bridge that gap,' wrote White. 'The medium's special gift or talent is the ability temporarily to step up her basic frequency – or allow it to be stepped up – to 150.'[2]

For the public, especially those who had personal experience of the military, this new metaphor – of the medium as the radio officer receiving reports from soldiers in the battlefield – only added to the popularity of Spiritualism. Most often, in fact, the messages received from mediums *were* from soldiers, and they were detailed enough to fill entire books which were published under the attribution of the soldier's name.

There are dozens of these posthumously written texts, many of which were directly influenced by *Raymond*. This is, again, where things become problematic. A cynic might argue that these 'authors' were leaping on the international commercial success of *Raymond*, copies of which were being sent to men on the Front as well as being bought up by those grieving back at home. It was an incredibly lucrative trend in publishing. But to read these books, at least the ones written by the family of a soldier killed in battle, is to see how genuinely they believed in the ghostly messages they received. It wasn't about money at all, but about preserving memory amid so much grief. No one's loss felt special in the grand scheme of things – the dead were lost in a sea of obituaries – but somehow everyone knew Raymond Lodge. Why shouldn't the world know about the life of their son, too?

In 1919, a woman named Mrs Kelway-Bamber got in touch with Sir Oliver Lodge, as so many hundreds must also have done following the publication of *Raymond*. She was wealthy and well-connected enough to have sittings of her own with Mrs Leonard – the medium who brought Raymond's messages to the Lodge family – whose schedule must have been rather full at that point. Mrs Leonard reached across the divide and got in touch with Claude, Mrs Kelway-Bamber's son who had eagerly joined the war effort in 1914 before he'd even had the chance to be conscripted. He was soon trained as a pilot and joined the Front as part of the Flying Corps. He lasted three months before he was shot down over Flanders, his plane crashing behind enemy lines.

In Mrs Kelway-Bamber's letter, she enclosed a manuscript for *Claude's Book* – a short record of the messages she had received from him through Mrs Leonard over 'many very happy hours.'[3] Initially describing herself as a sceptic, she said that it was only through her 'deep grief at his premature loss that she decided to investigate, in the faint hope that there might at least be some definite comfort in it.' Through reading books such as *Raymond* and some of the texts we looked at in the previous chapter, Kelway-Bamber became interested in mediums, automatic writing and the idyll of Summerland. She joined the London Spiritualist Alliance, attending the group's lectures and undergoing a few experimental sittings with mediums before managing to secure a place with the renowned Mrs Leonard.

Oliver Lodge's reply to *Claude's Book*, which Mrs Kelway-Bamber included – probably unbeknown to Lodge – as the foreword to the published text, was gently critical. It's clear he had reservations about some of the things Claude came out with, but to denounce them would be to denounce Mrs Leonard which would, of course, be to denounce Raymond's messages. There are some rather backhanded compliments which don't seem evocative of Lodge's usual enthusiasm for Spiritualist phenomena: 'I have read the type-script of your son's book, and though it may strike people as rather crude I am impressed by the honesty and simplicity and straightforwardness of its material.' He goes on to say that the book's content represents 'at worst a psychological phenomenon; while at best they convey the impressions of an eager newcomer to the other side' that, crucially, at times 'goes beyond actual experience and trespasses on the fanciful with too much of what is presumably hearsay and secondhand information – about reincarnation, for instance – all which for my part I discount.' It's interesting to note how he phrased this. Clearly, the messages Mrs Kelway-Bamber received troubled Lodge. While some are similar in content to the kinds of things Raymond described about Summerland and how the war is viewed from the spiritual plane, Claude does go on a ramble about reincarnation. Lodge does what we have seen Spiritualists do time and again in this book – there is always some sort of reason as to why an aberration in experience has occurred. In this case, Lodge claimed that Claude was overwhelmed by information in his new existence, and in discussing reincarnation was simply repeating myths and rumours that he had heard from his spirit-friends, who were probably just pulling his leg.

Claude described his moment of death as a 'terrible blow on my head, a sensation of dizziness and falling, and then nothing more.' He woke up in a hospital, a sort of triage for the newly dead in which nurses and doctors soothe the confusion of spirits who have just departed their bodies. We see again how different Claude's picture of the first moments of the afterlife is from other texts that seek to do the same – there is none of Stead's Blue Island, nor any mention of the 'Land of Utter Silence'. But there are similarities; Claude described each spirit having their own task to fulfil for the benefit of both the dead and the living, and each does their work gladly. He explained how he had been allotted a role of a science teacher. He again described the utopian, rural idyll, a rose-tinted version of the home from which he was so violently torn – there were 'beautiful woods and fields; the turf was springy, the air soft and clear, and soft sunshine over everything.'

As with *Raymond*, there is something about these texts, written in a time of heart-shattering grief, that makes them difficult to criticise. It problematises our ability to investigate these books with a sceptical eye. The same is true for *Claude's Book*, which Lodge clearly felt when he replied to Mrs Kelway-Bamber. There's a moment when Claude, having adjusted to his new existence and left the hospital of the afterlife, was permitted to visit his mother in spirit form. 'You were sitting up in bed,' he told her during one sitting with Mrs Leonard, 'in an agony of grief, the tears streaming down your face, repeating my name over and over again, and calling me, and saw me not.' There's a sense here of the pain so many bereaved families faced – not only were their sons dead, but they were dead over enemy lines, buried in foreign soil, out of reach in more ways than one. At the end of *Claude's Book*, his mother included a selection of his letters; they are presented so that the reader can investigate for themselves and compare the vocabulary and phrasing to the messages delivered in sittings, but really it seems as though they are to make Claude seem alive again, and remembered even by people who never met him. For Mrs Kelway-Bamber, Spiritualism brings her son within reach once more.

But it wasn't just the grieving British public who were fascinated by soldiers telling relatives about their death and the utopian life thereafter. A prime example of this Spiritualist trend reaching across the ocean is *War Letters from the Living Dead Man*, allegedly dictated to the American novelist Elsa Barker. It's a strange book that hints, I think, at Barker's general guilt at the United States' initial neutrality towards the Great War; yet, Barker has no direct link, is not directly grieving, and it makes the book somewhat doubtful in its intentions. She begins by outlining her psychic powers: she claimed she was a 'wide-awake astral participant in the first action in which the British army was engaged on the Continent' and that, while at home in New York, saw 'the shelling of Scarborough at the hour when it occurred.'[4] Then, one day, she was suddenly struck by the incoming signals of the soldier she calls 'X', and she rushed to begin writing down his extensive letters. The rest of the book contains his reports and updates from the other side, sometimes talking about specific events, but most often giving descriptions of life after death.

It gets very odd very quickly. On 17 March 1915, Barker's soldier said he'd met Archduke Franz Ferdinand, whose assassination in 1914 triggered the chain of events that led to the outbreak of the First World War. 'Yes,' he said, 'I spoke with him and advised him; but I had other things to do just then and left him with a priest of his own church, a gentle and strong soul who stood like a rock in the tumult.' A few days after this entry, he described how he and 20 other ghosts stood in the palace of Potsdam, 'trying by silent pressure of will to reduce the pressure of the war-will which surged in the German nation towards its emperor.'

He comes across as a rather audacious personality – I'm not sure why he deemed himself special enough to have a chat with the archduke, nor why he bore the responsibility of trying to reach through the ether to stop the conflict – and the rest of *War Letters* maintains an idea that 'X' was part of some sort of senior management team of the afterlife. He offered advice to prepare Barker's readers, too. Learn languages, he said. This doesn't feel consistent with everything else that 'X' described. In a form of existence where you can read the thoughts of the living, and travel from country to country in the blink of an eye, somehow for him language barriers still posed a problem.

These texts became a popular genre unto themselves, and it was mostly women who produced them. They feature a very similar structure and tone; the author invariably states that she is merely writing down the phantom words as they come to her mind. There is no doubt that *Raymond* influenced this surge of books describing death and the afterlife, supposedly from the direct correspondence of killed soldiers. And this phenomenon culminated in one particular book: Grace Garrett Durand's 1917 *Sir Oliver Lodge Is Right*. As with Barker's alleged discussions with 'X', Durand proclaims to have a gift for receiving and transcribing spirit messages as though through a wireless. She seems to have a penchant for famous ghosts, and lists among her correspondents Tolstoy, Joan of Arc and Abraham Lincoln (who asks her to call him 'Uncle Abe'). More unsettling than this, however, is that

Durand proclaimed to be in touch with one spirit in particular: Raymond Lodge. She wrote that Raymond was new to the spirit world when the book was written, and subsequently drifted across the ocean to the US to impart more information to a woman entirely unrelated and unknown to the family. There is something troubling, almost offensive, about Durand's book, particularly when she wrote:

> I have had several splendid talks with Raymond, and the conversation is the same as I would have with any bright, pure-minded, noble young man in the flesh. Raymond naturally is very much interested in the war and war conditions, and in a conversation with him a few days ago he spoke of our own American boys getting ready to go [to] the front, or of those who are already there, and said 'Mrs Durand, I love to look at them, they are such a fine lot of fellows.'[5]

She professed that she hadn't written to Oliver Lodge about these conversations, but Raymond allegedly encouraged her to publish a magazine article and send it to his father. There's no evidence whether she did so, or if the Lodge family ever read her book. Perhaps, in their inherent belief in Raymond's spirit, they would have been grateful that he was speaking to all who wanted to communicate with him. But there's just something distasteful about a complete stranger declaring that they regularly talk to the spirit of your loved one, without you knowing about it. In this way, interwar Spiritualism put private grief in the public domain; by turning the lost soldiers into larger-than-life characters, it encouraged believers to think that they had a right to carry out their own séances with other people's dead relatives. As we'll soon see, it was this sudden perversion of traditional bereavement that touched a nerve for sceptics.

Regardless of the controversy, these automatic-writing texts appear to have had a profound influence on readers, especially those struggling to come to terms with their own personal losses.

A sad example is that of a New Zealand woman, Elizabeth Stewart, whose faith in the veracity of Spiritualism led her to attempt to contact her dead husband. Lieutenant Colonel George Stewart died of dysentery in November 1915, in an Australian military hospital on the Greek island of Lemnos. Historian Rachel Patrick investigated Elizabeth's letters to her family and siblings-in-law, and charts her growing determination to speak to George using spiritualist methods. Elizabeth was active in the war, too, as a nurse stationed in Egypt. But once the war ended, she returned to the home she had shared with George, and that was where she believed she would finally reach him. In 1919, she wrote to a member of her family that she wasn't lonely in the house, as 'Geordie is very near & with this knowledge comes a great peace.'[6] While we may laugh at 'X' in conversation with Archduke Franz Ferdinand, or scorn Durand's alleged conversations with Raymond, to see an example of the personal impact of this cultural movement complicates the matter. The investigations into the afterlife, while often fruitless and disappointing, at least offered people a way to cope with their grief – they could at least feel as though they were *doing* something in the pursuit of messages from the dead.

As Lodge began to collect and corroborate evidence (Raymond described a photograph which the Lodges had never seen, but which is sent to them by a soldier sometime later), a new thread started to emerge. So far, our main characters were Lodge, his wife and Mrs Leonard, but Raymond also had two brothers, Alec and Lionel. They, as much as their parents, would have given anything to have Raymond back, but there was a clear distrust – more than mere scepticism – of Mrs Leonard's séances.

Alec and Lionel began to set traps. Secretly, they asked Raymond questions to which only the three siblings would know the answer. If Raymond really was with them, he would tap out the correct messages through Mrs Leonard. At the next sitting, which the brothers didn't attend, their questions were not addressed, and they told their father what they had done. Lodge, clearly flustered, had an explanation: it was the brothers' own fault. Had Alec and Lionel been there at the séance, 'it might have made the obtaining of the answers they wanted much more feasible, inasmuch as in their presence he would have been in their atmosphere and be more likely to remember their sort of surroundings.'

For every piece of speculative evidence, Lodge was delighted to demonstrate it as fact. For every hint at fakery, Lodge had an answer. So it goes on. *Raymond* is a very hopeful and optimistic book, but nevertheless I can't help wondering what we don't see, particularly in the family life of the Lodges. Alec and Lionel feature prominently in the book, each giving their own sceptical report. Yet, for all they don't believe that Raymond really had returned, they seemed to keep participating in sittings and experiments. Why? Was it to appease their father, who must have been pressing them to continue? Was it that they genuinely did cling to some sort of glimmer of hope that there was truth in it after all? Or was it, like Eleanor Sidgwick and Harry Houdini, that they wanted to expose the fraudulent medium before she could dupe any more people? I'm not sure it would be the latter, for the simple reason of Lady Lodge, Raymond's mother. While she is frequently present in the book, her words only feature on occasion. But when we do read her opinion on the matter, it is devastating.

'We can face Christmas now,' she wrote.

On one side, the book is bolstered by Lodge's stubborn determinism to describe the séances using scientific methodology; on the other, Alec and Lionel work in secret to diminish their father's results. At the book's tragic centre is Lady Lodge, grasping at whatever she can to ease the pain of losing her son. She is why the book is so complex and, at times, immensely difficult to read. As we also saw in the example of Elizabeth Stewart, these paranormal investigations could be a source of immense comfort on an individual level. Does scepticism and rationality still have a place in cases like these?

It's interesting, then, to find her quiet yet grief-torn presence also at the centre of the fierce debates that followed the publication of *Raymond*. One of the most prominent voices in the argument was Charles A. Mercier, Doctor of Psychology and Mental Diseases at the University of London. In the year following *Raymond*'s publication, and in the wake of its extraordinary and widespread success from the genteel parlours of middle-class America to the pockets of bloodied soldiers, Mercier wrote a scathing response. *Spiritualism and Sir Oliver Lodge* is a nearly 200-page assessment of Lodge's involvement in séance circles, and one that offers alternative explanations for what is presented as evidence in *Raymond*. He starts rather strangely, however. Mercier was sent a copy of *Raymond* and asked to give his professional opinion. He declared it was impossible for him to do so.

'The sorrow,' he wrote, 'of a bereaved mother is no fit matter for discussion by strangers in the public press.'[7]

I find this respect admirable, especially to outline it at the start of his book. I'm glad it's not just me who finds the most haunting figure of *Raymond* to be the young man's mother. But then it seemed that Mercier's seething rage for Lodge got the better of him, and he rather threw his initially delicate approach out of the window. He said it was impossible for him to review *Raymond* out of reverence for Lady Lodge, but on the following page he called it 'drivel'. He described Spiritualism as an epidemic. 'Like measles and scarlet fever,' he said, 'it never dies out.'

His main issue with Sir Oliver Lodge and Spiritualism appeared to be the scientific language with which it had recently been discussed. Mercier is quite clear that Spiritualism and anything relating to ghosts and the afterlife is not a science. He strikes me as a religious man, though, and on several occasions he outlined what he felt were the crucial differences between spiritualism and the Church. The latter was something 'on which the mind can repose in tranquillity'; he said it is up to the individual to interpret and appreciate religion in a way that meant something personal and private. He claimed that Spiritualists, on the other hand, cannot accept ambiguities: everything must be clearly and concretely defined as evidence in support of the afterlife, just as scientists conducted experiments in support of their hypotheses. It was this that offended Mercier.

He really didn't hold back in his analysis of *Raymond*, and some phrases are wince-inducing. For example, Mercier wrote, 'That Sir Oliver Lodge is a man who works at scientific subjects, and works at them successfully, is of course beyond all question; but there is all the difference in the world between a man who works at scientific subjects and a scientific man.'

Mercier emphasised that Spiritualism is not, and can never be, a science. And even if it were, what makes an astronomer like Camille Flammarion or a physicist like Sir Oliver Lodge qualified to investigate ghosts? The only type of person capable of truly understanding supernatural phenomena, according to Mercier, were conjurers. Only conjurers were equipped to separate tricks and fakery from what could potentially be genuine spirit communication. Harry Houdini would have been pleased by this endorsement. Enraged by the reports he read and by his own experience of fraudulent mediums, Houdini figured out many of the techniques to produce alleged paranormal phenomena – materialisation of objects, table-rapping, gloopy ectoplasm – and made it part of his act. Mercier made a fair point here, especially when the proliferation of exploitative con-artist mediums were using sleight-of-hand tricks to fool the hopeful and the grieving. It takes a conjuror to catch a conjuror, but also to be able to admit when phenomena truly have no explanation.

What I find difficult about *Spiritualism and Sir Oliver Lodge* is the unbridled vitriol with which Mercier wrote. At one point, he even left Lodge alone for a moment to target Sir Arthur Conan Doyle instead with this biting criticism:

> If Sir Conan Doyle understood what scientific method means, and what the spirit of science is, he would be ashamed to adduce authority at all in support of that which ought to rest on evidence and reason alone; and if he understood what may and what may not be accepted on authority, he would be ashamed to adduce, on the ground that they are not well known, the names of men as authorities in a subject of which they have no special qualifications for judging.

Doyle lost a great deal of his family in the war. First, his brother-in-law Malcolm Leckie was killed at Mons in 1914. Then, Doyle's son died in a London hospital in 1918 after being wounded in the Somme, and was shortly followed by Doyle's brother Innes. A close family friend, Lily Loder-Symonds, lost three of her brothers in 1915. According to Jay Winter, as a Spiritualist and self-proclaimed medium Lily was 'fundamental' in converting Doyle from a staunch sceptic into one of the most outspoken advocates for communicating with ghosts.[8] After suffering such intense and repeated bereavement, and witnessing the grief of people close to him, it's understandable that he would want to cling to any hope of being reunited with his loved ones. So when Mercier cut so deeply into Doyle's belief, it feels unnecessarily cruel. If Mercier can spare some sympathy for Lady Lodge, why can't he do the same for the grieving fathers too? It seems to be about professional integrity. Mercier couldn't abide Spiritualism being investigated in a scientific manner to search for proof – only to expose fakery. Mind you, he was also infamous for declaring vegetarianism as a symptom of insanity, so he clearly wasn't someone who could see things from another person's perspective. If I've learnt anything from reading about these wartime investigations, it's that empathy is crucial.

But that's not to say that critics such as Mercier and Houdini were completely out of order in dashing the hopes of the grieving public. As much as publications celebrated the alleged return of killed soldiers, fraudulent mediums saw a business opportunity in the suffering of those left behind in the war. Where Lodge and Mercier argued their case through the writing of books, there were some who took a far more active approach in the exposure of con artists in the séance room.

Surprisingly, I didn't have to look very far for an example of this game of cat-and-mouse. The archive of my local newspaper, the *Cambrian News*, featured a lengthy report in May 1918 of a melodramatic case from my home of Aberystwyth.[9] During this time, fortune telling was an arrestable offence under the Vagrancy Act, and so clairvoyants practised a secret, illicit trade that was nonetheless still widely popular. The war made them even more so, as not only were people anxious to contact their lost boys, but they were also keen to know of any sign of how their own lives might become tangled in the conflict – and whether they were soon to be conscripted themselves.

Madame Smith Evans, also known as Madam May, offered just such services. Aberystwyth's police force were having none of it.

On 15 April, late in the evening, Police Sergeant Daniel Thomas set his trap by dressing as an army major and knocked on the door of Madam May's townhouse. He was led inside by an elderly maid, where Madam May looked at him – clearly eyeing his uniform – then 'shuddered and sighed'. She began to have convulsions, a fit of coughing, and finished up by vomiting.

'You have been gassed,' she told Thomas.

She complained of pains, of the pains she said Thomas had himself suffered. Dysentery, an illness of the bladder and mental anguish were all in Thomas's past. Then she suddenly covered her ears in horror.

'Oh, what terrible noise. Have you had shell shock?'

I wonder what would give her that idea.

Madam May had a co-conspirator in the room, a Mr Evans, who added flourishes to this scene but otherwise let her take centre stage. Now Thomas was told by Evans that he would be called back to the Front, suffer greatly, but would survive and be all right. Half a crown, please.

Four days later, the police had another go, this time sending a police constable disguised as a collier. Again, Madam May received him with a similarly theatrical display; she complained of pains in her back, her head, behind her eyes, and said she was dreadfully cold. Mr Evans joined the two, and between them – using the constable's gloves as a kind of spiritual token – discussed his fate. The constable hadn't even asked, but Madam May and her confederate assumed that he had come to learn about whether he would be sent to war. No, they decided. Being a collier with presumably blackened lungs, he would be judged too sick and weak to be conscripted. Half a crown, please.

Superintendent Phillips then visited the address on 27 April, and summoned both Madam May and Mr Evans to court. They pleaded they didn't read fortunes but were practitioners of psychometry, and compared themselves to prominent psychical researchers such as Sir Oliver Lodge. Madam May and Mr Evans were both fined, and reported to the War Office. One of those present in court told them that the 'uniform has been insulted'.

It was through criminal cases such as this that spiritualism became a heated political issue in the years following the First World War. The Vagrancy Act of 1824 was synonymous with a kind of witch trial – indeed the 1735 Witchcraft Act was also still in effect at this time – and the growing ranks of Spiritualists began to campaign against the persecution of mediums. In 1916, for example, the American medium Almira Brockway was invited to England to demonstrate her powers, but was arrested soon after arriving for no particular reason other than that she claimed to speak to ghosts.[10]

Between the wars, parliament repeatedly engaged with the issue of Spiritualism, and it was the Labour Party who sought to amend the laws against the practice. In 1930, Sir Oliver Lodge and a group of Spiritualists were invited to 11 Downing Street by the Labour chancellor's wife, Ethel Snowden. There, they discussed the importance of public acceptance on the realities of communication with the dead. The same year, Labour MP William Kelly tried to pass the Spiritualism and Psychical Research (Exemption) Bill to stop séances being disrupted by overzealous police officers. Only 40 MPs attended the debate in parliament, however, and the Conservative MP Francis Freemantle, comparing the issue to the nonsense of fairy stories, saw to it that the Bill was rejected.[11] The Witchcraft Act and its associated laws against mediums was finally repealed in 1951.

A similar situation was simultaneously occurring in New Zealand, whose Spiritualist believers had grown exponentially over the previous decades. In order to band against a sceptical government and strong resistance from high-ranking members of the Christian church, an organised body called the National Association of Spiritualist Churches of New Zealand was formed, and remarkably managed to pass a Bill in 1924 that protected them from some of the persecution affecting their British colleagues.[12] However, mediums were still arrested under an outdated British Crimes Act which included sections on witchcraft. Indeed, New Zealand's laws against witchcraft weren't fully repealed until 1981, where the fortune-telling section was replaced with a law against mediums who were clearly intending to deceive their paying customers.

Where nineteenth-century spiritualism was, at worst, dismissed as a foolish pastime, it was against the backdrop of widespread grief and the open wounds of the First World War that it became an increasingly controversial part of society. Perhaps, as we've seen, it was because it suddenly offered so much hope to so many people, but it equally provided golden opportunities for exploitation. It's a very difficult and murky period in the history of paranormal investigations. On the one hand, it exposed a grim trend of fraudulent mediums preying on the public's grief. Yet, it clearly did give comfort to people such as Elizabeth Stewart and Lady Lodge, and the value of that can't be easily dismissed.

While the effects of the Second World War didn't have quite the same influence over Spiritualism as the first, the recurrence of mass bereavement still encouraged people to seek solace from the possibility of spiritual survival. In Chapter 2, we met Sir Oliver Lodge's Welsh assistant, Benjamin Davies, who worked with him both on matters to do with Lodge's laboratory and his interest in psychical research. Steadily, over many decades, Lodge encouraged Davies to become increasingly swayed by Spiritualism. Davies, to begin with, approached it from a methodical, scientific perspective. He invented electrical equipment to test the veracity of mediums, was sent on missions to investigate mediums and poltergeist cases by Lodge and the SPR, and kept fastidious records in separate notebooks of his experiences of séances. One such notebook containing Davies' ephemera from his friendship with Lodge included the following letter from Lodge: 'Raymond is killed. Telegram from war office came on Friday.'[13] It was, presumably, Davies who cut out and stapled to the letter the small announcement in the *Chronicle* dated 20 September 1915. On the following page, Davies kept the memorial to Raymond – perhaps he attended a service – on the back of which is written, 'His Parents and Family, well knowing that thousands of others are similarly stricken, are grateful to the many friends who have sympathised in their bereavement.' For Davies to have kept these tokens of Raymond's untimely and violent end, he must have felt the loss keenly.

That loss was felt again at the start of the Second World War in August 1940, when Davies received a letter from Lodge's secretary, Helen Alvey, to inform of his mentor's death. 'Sir Oliver went most happily and peacefully,' she wrote, 'and we think of the welcome he is now having on the other side.'[14] Davies wrote back lamenting the effect the war was having on delaying the post. In his reply, it's clear to see how Davies' involvement in Spiritualism had developed from his early days of visiting mediums on Lodge's behalf. He wrote, 'I am glad to realise from your letter that you possess Sir Oliver's view of death and the life beyond; and I hope that members of his family share it. [...] The welcome to Sir Oliver must be a notable one, for he was a mighty leader in the great movement.'[15] No doubt both Helen Alvey and Davies, in their sentiments about Lodge's welcome, were thinking of Raymond as they wrote it.

After Lodge's death, while the Second World War raged without him, Davies went on one final mission in line with the kinds of tasks his old mentor used to set him. Back in Aberystwyth, the place where he had investigated the dud poltergeist on Lodge's and the SPR's behalf, Davies heard of the death of Eric Evans, the son of one of his friends, Mr Ivor Evans, and wrote to him. He waited a little while after Eric's death was announced to contact the Evans family, firstly because he felt a letter from him wouldn't have done much to help their grief in those early days, but also because he wanted to allow for their shock to subside to have the 'opportunity of saying or writing a word more effectively.'[16] In other words, he wanted to discuss Spiritualism with them. He told Mr and Mrs Evans that 'Eric is still with you and, in all probability, spiritually nearer than ever before, for love on the home hearth was strong. Such love endures, and is largely the means of communion between the two worlds.'

In convincing them of the veracity of Spiritualism, Davies offered a summary of his own conversion. 'I began to realise this important fact in my younger days,' he wrote, referencing the work he carried out for Lodge, and how Lodge's mentorship encouraged him to explore Spiritualism with something other than through a purely scientific lens. 'The realisation', he continued, 'led me to join the Society for Psychical Research' – during the time when Arthur Balfour, Eleanor Sidgwick's brother and future prime minister, was president. He described that decision as one of the most important he had ever made. To the Evans family, he described a faith that seemed influenced by notions of Summerland: 'There followed a quest which is still continuing. Gradually, the future beyond the veil lifted itself. The universe became more and more spiritual. Death's sting was destroyed. The life course of a good soul through the series of ever ascending states of man's existence became clearer. A finite being is then given a vision of purpose and grandeur of the father's universe.' Interestingly, he gives Summerland, 'the next states of being', a Welsh name: *Trigfannau ty fy nhad* – the dwellings of my father's house.

A month later, still anxious to help the Evans, who were starting to get wrapped up in Davies' promises of life after death, he wrote to the London Spiritualist Alliance. This, I think, is fascinating, and shows how Davies' belief had developed; the SPR, after all, were still sceptical of mediums, but Davies was now very much converted. It's a far cry from his earlier notebooks, where he drew up plans to mildly electrocute mediums to prove whether or not their powers were genuine. Now, in the midst of the Second World War, he had no doubt. The SPR wouldn't help him convert Mr and Mrs Evans, so he turned to the LSA instead. He wrote to their secretary, Mercy Phillimore, an influential journalist who was extremely well connected in Spiritualist circles, asking for advice on how best to help Mr and Mrs Evans get in touch with Eric. To Phillimore, he wrote: 'Could you advise me as to the best medium for his purpose and whether something could be done by proxy sittings to begin with? I mention the proxy sittings because my friend is a busy man.' Indeed, Mr Evans was a well-known and successful solicitor in Aberystwyth, with a large number of staff working for him. But because he had others to take care of business, Davies suggested that Mr Evans could travel to London. As we saw during the police rumble a few decades earlier, Aberystwyth wasn't exactly known for its successful mediums – all the celebrities whose feats were proudly written about in the Spiritualist journals were down in England's capital. He even mentioned Mrs Leonard, who put Oliver Lodge in touch with Raymond, on a hopeful note, despite acknowledging that she no longer practised mediumship. Phillimore promised Davies she would do all she could to help his friends.

Mr and Mrs Evans were soon hooked on sittings, frequently travelling to London to visit a medium called Mrs Nash. In June 1944, Mrs Evans enthusiastically wrote to Davies to thank him for encouraging them to seek contact with Eric, and had been comforted in their grief by a particular message: 'Tell father I am alive.'

As the public began to heal from what they had lost in the Great War, and more mediums were frightened away from their practice under the threat of the Vagrancy Act, interest in séances and automatic writing gradually faded. Nevertheless, ghosts remained as popular as ever. With every major cultural event or movement, paranormal investigations changed to suit the public mood. The Second World War did not feature a repeat of the fervour for spirit communication. The interwar period to the middle of the twentieth century saw parapsychology develop into a discipline, and evidence was examined under laboratory conditions. But the supernatural activity itself changed, too. Indeed, as bombs began to shake the cities of allied countries, ghosts became violent. The poignant voice of the young dead soldier was replaced by the outburst of the poltergeist.

CHAPTER SEVEN

The Ghost Laboratory

In 2022, a team of psychologists from New Zealand created a ghost in their laboratory.[1] They wanted to know whether people would report anomalous experiences in a particular location simply because they'd been told that a death had occurred there, and that the environment might now be haunted. They recruited 100 participants, who were randomly assigned to a control or experiment group. In the control group, the participants simply had to wait in a room. Those in the experiment group, however, were told that a (fictitious) caretaker had recently died in the room and that it was policy to inform them. While closing the door, the researchers casually explained that a PhD student had sworn they'd experienced something weird. The participants were asked to conduct a mindfulness task, focusing on their breath and remaining calm. There was also a lamp in the room which, unbeknown to any participants regardless of whether they were in experiment or control conditions, could be remotely switched on and off using Bluetooth.

Three minutes in, the lamp suddenly turned off for exactly seven seconds, plunging the room into darkness before turning back on again.

The researchers found that regardless of whether the participants believed in ghosts or not, the non-control group were markedly more distressed by the sudden blackout than those in the control condition who hadn't been told of the death or possible haunting. By creating a ghost in their laboratory, the researchers showed that, actually, sometimes it's not about whether we are believers or sceptics of the paranormal: there's almost something ingrained within us to be distressed by unexpected occurrences when we've been told that a location is haunted. I know I certainly would have bolted from the room when the light went out, because I've come to understand that my own scepticism is a fragile thing, and perhaps it's more of a front I put up to make myself feel braver in dark and creepy places.

This laboratory experiment is a recent example of a line of inquiry into the paranormal that is now over 100 years old. After the establishment of psychical research societies and ghost-hunting clubs, investigations into ghosts went a step further, and tried to institutionalise experiments within properly equipped and maintained research environments.

It all began with Harry Price. For better or worse, he appears again and again in the history of ghost hunting because he had such a profound influence on twentieth-century practice. As we saw during the discussion of Borley Rectory, in the 1920s he broke away from the Society for Psychical Research to establish his career on his own. He wanted the fame and the glory for himself, and he couldn't do that while tied to a society that was collectively more prominent than he was – the publicity would always go to them. He decided to begin conducting his own experiments and investigations, in controlled environments such as the laboratories scientists were increasingly using.

In 1925, he published an account of an investigation which was a precursor to his next step of establishing his own ghost laboratory. It was an investigation into a 23-year-old London hospital assistant and medium named Stella Cranshaw (or Stella C. as she is known in the book).[2] During her séances, the usual raps and lights occurred, but objects also appeared to move around her, and, interestingly, she seemed to be regularly startled by her own phenomena. Price began to have sittings with her in 1923, and immediately took steps to control and equip the room in the same manner as a rigorous scientific experiment. As we saw with Borley Rectory, Price had no shortage of money to burn on his ghost-hunting hobby, and the same was true for the Stella C. experiments. In one corner of the room, he fastened to the wall a state-of-the-art thermometer made by renowned scientific and optical equipment company Negretti and Zambra. He used red lamps to illuminate the séance room, and red pocket torches allowed the investigators to take scrupulous notes and draw minutely detailed plans of the room. Temperature readings were taken regularly.

What Price found was that in the moments before a 'violent manifestation' from Stella, the temperature dropped. By recording these drops in temperature, he showed that some physiological phenomena – such as feeling a sudden, ominous, cold breeze – could be objectively measured and proven as having really taken place in the room rather than in the imagination of those present. Sure enough, each time such an experience was had by the group, something dramatic would occur: the table 'reared on two legs, and rapidly moved across the room' on one occasion. It's reminiscent of the Philip Experiment we saw in Chapter 1; before the phenomena began, the Toronto team of researchers always reported a feeling of rising tension around the room. Indeed, at one point in the Philip Experiment, the phenomena were so powerful that the table broke; during a sitting with Stella C., the table top snapped in half without warning, and the legs were reduced to 'matchwood'.

Price's séances with Stella C. gained him further notoriety, and the year after the book's publication, in 1926, he established his National Laboratory of Psychical Research. The laboratory was fairly short-lived, dissolving completely in 1938 after it had been turned into the University of London Council for Psychical Investigation, but not before it had been used in one of ghost-hunting history's most dramatic exposures of a medium.

In the late 1920s and early 1930s, Perthshire-born Helen Duncan was a hugely popular and successful medium in the UK. Her powers were rather like those of Eva C.; she could materialise spirits and long, flowing white streams of ectoplasm that seemed to spill like milk from her nose and mouth. She was chronically ill, had tempestuous mood swings, and by the age of 32 had six children – a further three having died in infancy. She had, on several occasions, been suspected of fraud, but Price invited her to his laboratory to undertake his own experiments.

Helen travelled across the country to give sittings with her husband Henry as her carer and manager. In late 1930, Price heard that they were due to be in London soon, and made an offer to employ Helen's services for several months while he conducted his tests.[3] It took a while for Helen to be persuaded, but by early 1931 the Duncans agreed to give Price and the other members of his laboratory a private sitting. 'This was not a rigid test,' Price recalled, 'merely a friendly demonstration under laboratory conditions in a properly equipped séance room.'

Immediately, the experiments began with a wincingly thorough examination of Helen Duncan. Previously, it was suspected by other sceptics that she had, like Eva C., hidden her 'ectoplasm', this time suspected to be cheesecloth rather than paper, in compartments of her clothes or in various orifices. Inside Price's office, and, allegedly, at her own suggestion, Helen stripped naked and was examined by his long-suffering secretary Mollie Goldney. She looked through Helen's hair, down her throat and up her nostrils, but did not, for the sake of keeping the mood friendly, conduct a vaginal examination.

Satisfied that there was nothing concealed in the places they had checked, the laboratory members made Helen wear the thoroughly unflattering 'test' garment the Duncans had brought: a large pair of black trousers, stockings and a coat, which were searched and then fitted to Helen so that she couldn't conceal anything new within the folds of the fabric.

As with her predecessors, Helen used a séance cabinet in her sittings. Price had assembled one in a chimney recess of his laboratory, consisting of an armchair that could be hidden from view with plush red curtains. After Helen had changed into her medium's garb, the assembled researchers and Mr Duncan led her to the chair and closed the curtains. Price remarked that 'in 15 seconds the medium was apparently asleep and quietly snoring.'

Price had thought of everything in his laboratory. He had even kitted out the floor to provide the best results for his ghost-hunting experiments, using cork for the flooring to maintain a good temperature. All manner of lighting was available, from red and other coloured bulbs to torches, lanterns, candles and bright floodlights, and he had even forked out for ultra-violet and X-ray apparatus. Elsewhere in the room were Dictaphones, cameras, stopwatches, thermometers, rheostats, barographs and all manner of equipment to record every possible variable within the room.

With Helen in the cabinet, the assembled party waited. After a little while, the curtains slowly drifted open, revealing Helen in a trance state. Bright white ectoplasm, or 'teleplasm' as Price called it, coiled like a worm from her mouth all the way down to her lap. The curtains closed.

When they opened again a few minutes later, the ectoplasm had spread, and was beginning to trail along the floor. Then it spread even farther; the next time the curtains opened, it was 'like an apron,' then like a 'bridal veil', now about 2.4m (8ft) in length. It was a curious substance, filmy and translucent.

The séance went on like this for a while – the ectoplasm spread, receded, wrapped itself around Helen's arm, squirmed along the floor, while the curtains opened and closed again, each time revealing a new spooky tableau. At one point, Price reached down and picked up a tendril of ectoplasm from the floor and did what any self-respecting investigator would do: he gave it a little sniff. To his faint disgust, it had a strange, musty smell, and was damp to the touch.

They then asked Helen's spirit guide, a cantankerous personality who called itself Albert, if they could tie Helen to the chair. Albert had been a part of Helen's sittings for many years, and was alleged to speak through Helen via ectoplasmic lips. Albert, in his gruff voice, permitted this to go ahead. However, as soon as adhesive tape was used to strap Helen to the chair, the ectoplasm ceased to flow.

Price took stock of the sitting, reading back over the notes hurriedly made during the course of the phenomena. Everyone seemed to be amazed at the way in which the ectoplasm flowed from Helen, but now, looking back, Price wondered if they did see that at all, or whether it simply *looked* as though it were issuing from her nose and mouth but had, actually, just been tucked into her orifices.

He pondered, moreover, at the moment in which he held – and, lest we forget, sniffed – the ectoplasm. It didn't sit right with him. It wasn't as ethereal as he'd imagined; it felt 'just as if I had caught hold of a man's vest'. The ectoplasm, allegedly the stuff of spirits themselves, felt awfully woven in texture, as though it were simply a strip of material. The smell, too, was less than paranormal, and seemed more in line with the odour of human saliva.

However, Price was still inclined to give Helen the benefit of the doubt. He hadn't, at that point, cut off a piece of the ectoplasm. As we've seen before, Spiritualists had created an entire mythology around materialisation and the fragile link between a medium and any kind of ectoplasmic substance. When William Volckman seized Katie King, her medium Florence Cook claimed she had been 'burned' by the touch, feeling King's pain amplified on her own skin. This idea had continued well into the twentieth century, where mediums claimed that to touch or damage their materialisations in any way could cause them serious injury. Price was clearly taken with this theory, and claimed that the reason he hadn't snipped off a bit of Helen's 'teleplasm' equated to a doctor not suddenly snatching a patient's heart out of their chest while they lay on the operating table. In any case, he said, the 'crude art of "grabbing" is not countenanced by the laboratory officials'. If he succeeded in retrieving a piece of ectoplasm, it would be because Helen (or Albert) gave it to him.

Nevertheless, his scepticism rumbled as Helen participated in more laboratory séances. During a séance in May, the usual phenomena occurred, and the cabinet curtains parted to reveal Helen, seated in the armchair, with a stream of ectoplasm falling from her mouth. This time Price decided, despite his own rule, to do a little gentle grabbing. He reached out to poke his head into the cabinet, so he could see what happened when the curtains were drawn. As soon as he moved, however, Helen shifted as well.

In the quiet of the laboratory, a faint, metallic clink was heard.

It sounded to Price as though the ectoplasm had been pinned to the curtain, and Helen's sudden movement in reaction to Price's outstretched hand had unfastened the safety pin. Mollie Goldney found the pin attached to the curtain, and recognised it as the one she used to attach Helen's séance coat together.

Having read many of Mollie Goldney's letters and reports, I feel as though I know her quite well. She was an incredibly sceptical ghost hunter, and I highly doubt that at the sight of her safety pin in the curtain, she would have even entertained the notion that it got there through paranormal means. Quite a lot of her letters to Eric Dingwall demonstrate her frustration with Price, and I suspect this was the case here, too. Price, rather than suspecting Helen of fraud, said that the pin incident only proved that the ectoplasm was strong enough to open a safety pin when pulled, and that such an action did not, in fact, hurt Helen.

Nevertheless, he was troubled by this development. Perhaps Goldney had eventually talked him into seeing the matter as she did, but after a while he began to conclude that there wasn't anything ghostly about Helen's materialisation. The ectoplasm was clearly made of some kind of material, most likely thin cheesecloth that could be folded and rolled up to a very small size. The question now, however, was how Helen was smuggling these lengths of cloth into the laboratory.

And that was where things started to go wrong for everyone involved in Helen's laboratory sittings. Price felt compelled to find the answer, to prove just where Helen's ectoplasm was hiding when she entered and left the séance room each session. Earlier, we saw how, out of so-called courtesy for Helen, the laboratory investigators didn't conduct a vaginal examination prior to the séance. But Price had no qualms about that now, and it seems that Helen's husband decided on her behalf that she would willingly submit to such humiliation. The description of this procedure is quite disturbing. Mollie Goldney brought Helen into the usual room in the laboratory, and the doors were locked. Then Mollie (who worked as a midwife alongside her ghost hunting) and Dr William Brown, another researcher involved in the laboratory, began their examination. Horribly, Price wrote in detail about how 'the rectum was examined for some distance up the alimentary canal and a very thorough vaginal examination given'. They didn't find anything. Price ended this entry in the report by thanking Helen for her cheerful cooperation. This must surely be an exaggeration.

Still, the phenomena continued. Ectoplasm issued like never before from Helen's nose and mouth, flowing like tendrils of smoke down her chest and onto the floor. Price had another theory, though: Helen was swallowing little bundles of cheesecloth and regurgitating them inside the cabinet. The more he suspected, the more he began to see evidence. Taking photographs of the brief moments in which the curtains parted and Helen's ectoplasm is revealed, he noticed new details, such as how the ethereal stuff – allegedly the very material of the spirits themselves – had a manufactured hem.

As mentioned earlier, Price's National Laboratory of Psychical Research was kitted out with some very expensive and advanced equipment, including an X-ray machine. The investigative team decided to find out whether Helen had been swallowing the cheesecloth. Once more, Helen's husband agreed on her behalf, and said she was keen to undergo an X-ray of her chest and stomach. Several doctors and scientists gathered in the room to operate the machine, and a few other members of the laboratory stayed to witness the examination. Helen sat on the sofa in the room, smoking one of Price's cigarettes. She watched as the men set up the machine, eyes darting between them and the strange apparatus.

And then, Helen Duncan bolted. She declared she would not undergo the examination. Her husband, Henry, tried to calm her down, to reassure her that there was nothing to worry about, as did everyone else in the room. But this only made things worse. Helen leapt out of her seat and whacked Henry across the face, then went for Dr William Brown – the same man who had a few days prior performed the humiliating genital examination – who managed to dodge a similarly powerful blow. Helen was by no means a diminutive woman, and was out to cause serious pain. A little more subdued, she then demanded to be taken to the lavatory, and Goldney and Dr Brown offered to accompany her down the corridor to the toilets. Suddenly, her anger seemed to peak again, and she shoved poor old Mollie Goldney out of the way and ran out into the street. Her husband soon joined her outside to calm her down, and half an hour later she was persuaded to return – but her séance garment was now torn open in several places. Mrs Duncan, much to the laboratory members' surprise, now demanded to be X-rayed, and nothing was found.

At this point, the laboratory's council were seriously discussing the veracity of Helen Duncan's teleplasmic powers. They no longer believed her to be genuine; and the tearing of the séance dress while Helen was outside suggested that she and her husband were busy discarding evidence before she consented to be X-rayed. It was clear to the members now that Helen was in the business of concealing the teleplasm about her person, whether managing to keep them somehow inside the séance garment or in her throat or stomach until needed.

The laboratory members' next task, after the failure of the X-ray, was to acquire a sample of her teleplasm that could be analysed, and during a later sitting they all sat there brandishing scissors rather menacingly. Helen's spirit control, Albert, consented to the 'operation', and a researcher Price refers to as Dr X prepared to snip off a piece of the gloopy white material. However, each time he tried to close the scissors, Helen moved in a seemingly accidental, jerky manner. Dr X held the teleplasm firmly, and as Helen suddenly twitched her head away again, a piece tore off in his hand 'exactly like tearing a wad of damp paper'. Helen screamed as though a limb had been ripped from her body, perhaps to distract the laboratory members and cause them to discard the teleplasm out of concern. But Dr X held on to the fragment, and upon analysis found it to be 'exactly like a sheet of paper which had been rolled into a tube and then flattened out' in a pattern which would make it easy to swallow. Note here that Price repetitiously used the phrase 'exactly like' – he still wouldn't go as far as to say that the teleplasm *was* paper.

This lingering hesitancy to fully admit to the truth was abandoned on further chemical analysis. Price admitted finally that it was, in fact, paper. His colleague at the laboratory, Dr Fraser-Harris, prepared the sample for inspection under a microscope, and the fibres of the paper were clear to see. They made their own control slides using blotting paper and filter paper to compare, and found them both to be indistinguishable from Helen's teleplasm.

Helen's teleplasm had varieties, however. There were the long, trailing strips the laboratory members knew to be cheesecloth, and the flat, fibrous type which had been proven to be made of paper. But there was also a third type – gelatinous, gloopy, mobile. Price managed to take a sample of this, trapping it in a bottle. To his surprise and probable disgust, after a few days he found that the blob of ectoplasm was turning black and going mouldy. It was rotting. He sent it to Dr X, whom I'm sure was delighted to receive it in the post. After examining the sample and conducting some chemical tests, Dr X found that, simply, it was cooked egg white mixed with a few other ingredients to keep it in one piece.

If you'd like to amuse your friends and family by regurgitating your own teleplasm, Price included a handy recipe:

> White of a new-laid egg,
> Ferric chloride,
> Phosphoric acid and stale urine thickened with Nelson's gelatine,
> Hot margaric acid from olive oil.

He instructs to pour the mixture into a tube which can be sealed at either end and submerged into boiling water for 15 minutes. Once it has cooked, you will have a 'beautiful strip' of teleplasm.

While Price's exposé of Helen Duncan's materialisation trickery didn't spell the end of her career, it followed her like a dark shadow, and she was desperate to stay in business. In 1941, during a séance, Helen revealed that a British Navy ship called the HMS *Barham*, crucial in the war effort, had been sunk – a fact which was being kept under wraps by the Navy. This quickly aroused the Navy's suspicion, who caught wind of Helen's latest clairvoyant phenomena. It was widely known by this point that Helen had been repeatedly exposed for fraud, and the Navy didn't for a moment believe the dead sailors of the *Barham* had been speaking to Helen. There had obviously been a leak, and Helen was now culpable in unwittingly spreading official government secrets. The Navy kept an eye on her, and by 1944 two undercover lieutenants had seen enough of her fraudulent séances and reported her to the police. She was arrested, as so many other mediums had been, under the Vagrancy Act – but later the charge became that of the Witchcraft Act, making her the final person to be charged and imprisoned under this act, much to the fascination of the public and the press.

Among those most invested in the Duncan case was Mollie Goldney, whose letters to Harry Price during the trial are unusually imploring of him. Having witnessed Helen's 'powers' in the laboratory, she knew with absolute conviction that they had obtained enough proof to help the prosecution and to stop the general public from seeing her as something of a martyr – which is what the Spiritualists were trying to do. Price's report on Helen had been published in 1931, and Goldney begged him to rewrite and republish it into a book as commercially successful as 1940's *The Most Haunted House in England*. She wouldn't let it go even after Helen had been convicted and imprisoned. In one letter, she thanked Price for copies of the 'preposterous' photographs of Helen during their 1930 laboratory experiments, and replied: 'But WHY did you not give these to Prosecuting Counsel so that they could be handed to the 50 witnesses for their comments?! It would surely have stopped the witness in her favour more effectively than anything else?'[4] Shortly after, Price's laboratory disintegrated, and nothing of its specific kind was ever re-established.

Being a laboratory-based professional ghost hunter also meant doing extensive fieldwork. High-profile cases of hauntings were increasingly in the news, and psychical researchers were adapting their roles to include aspects of medicine, archaeology, physics, chemistry and forensic detective work. And each of these aspects come with their own set of specialist equipment: a doctor needs a stethoscope, an archaeologist a chisel and brush, and a detective a magnifying glass. Eric Dingwall recognised this, and designed his own ghost-hunting kit.

The kit belongs (ironically) to the Harry Price archive at Senate House Library in London, and when I visited I got the chance to see it. I waited at a desk in the small reading room as a curator went to the store room, and brought back a grey archive box. Carefully, he opened the lid and the sides of the grey box fell apart like a flower bud to reveal the crumbling red kit within. I was surprised by how compact and unassuming it was. When new, it would have been a smart little red case, smaller than a shoe box. If Dingwall walked around with it, no one would have been able to guess at what lay within. The curator, with only the forefinger and thumb of each hand, lifted away the top – which, once attached to the rest of the kit, had fallen off with age – and placed it gingerly on the desk.

'You can take things out of the box to look at,' he whispered.

'*Really?!*'

'Yes, it's only the box that's fragile.'

Even so, the contents of the box were now around 100 years old. And besides, by this point I had read so much about Eric Dingwall, Mollie Goldney and Harry Price that something as significant to that part of history as Dingwall's ghost-hunting kit seemed like a priceless relic I shouldn't be allowed to touch anyway. I took a few photographs of what I could immediately see, but it was soon apparent that various items of equipment were hidden under what Dingwall had crammed into his little kit. I delicately, reverently, took things out to explore the kit in more detail.

What does a ghost hunter need to catch his elusive prey? Surprisingly, the contents were very simple. A brush was strapped to the lower flap of the box, and snuggly sitting in the top were several bottles and phials of powders and chemicals including petrol and turps. In a thin red Senoritas cigarette tin were a number of pins, the heads of which were covered in a chalky-white substance which would have once glowed with phosphorous. There were a number of matchboxes, clumps of cotton wool, a tape measure, half a stick of sealing wax, spools and tangled balls of thread, string and wire, and two compasses.

One of the most interesting things about the kit, however, was not its contents at all. Attached to the inside of the broken-off lid is a handwritten note from Dingwall himself, written upon presentation of the kit to the library. It reads:

> This is E. J. Dingwall's box of necessities for haunting and poltergeist investigations, an idea later borrowed by Harry Price who pretended it was his idea. See his 'Confessions of a Ghost Hunter' (London, 1936), p. 32 with photo. If the University of London would like it as a memento of the donor of the indexes please let them have it. E. J. D.[5]

He couldn't resist one last dig at his rival.

Also in the library's archives are a number of original photographs of Harry Price's kit, taken by Price himself to include in the book mentioned by Dingwall. The difference in equipment is startling. Where Dingwall's kit was small, unassuming, full of bits and bobs you'd find in a garden shed or a trouser pocket, Price's is the exact opposite. In the middle of the photograph is a large, smart-looking briefcase on and around which is piled so much expensive equipment it's a wonder how any of it fitted inside. And most of the kit's contents were, for the time, state-of-the-art technology. There is a camera and video recorder, electric torch, various light bulbs, an electric bell, and a handsome box of assorted callipers, protractors and other finely made drawing and measuring equipment. It's almost as though Price is saying that a *real* ghost hunter doesn't faff around with string and cotton wool and luminous pins. His impressive collection of equipment demonstrates his wealth, his tendency for one-upmanship, but also his need for publicity. There's a sense that Dingwall's kit is a personal, private little toolbox, where each item is there to serve a simple function without drawing attention to itself, whereas Price's kit exists for other people to see it and to ask about the function and purpose of each device.

In his 1940 book about Borley Rectory, *The Most Haunted House in England*, Price describes what he took with him to investigate the ghostly phenomena, and it includes far more equipment than in the photograph. His list features, among many other items, felt overshoes, a steel measuring tape, bits and pieces to fasten windows, doors, cupboards and seal draughts, wire, batteries and other electrical equipment, a camera and bulbs, a telephone, a bowl of mercury 'for detecting tremors in a room or passage or for making silent electrical mercury switches', notebooks and pencils, a flask of brandy and iodine (a ghost hunter's first aid), a 'sensitive transmitting thermograph, with charts, to measure the slightest variation in temperature in supposed haunted rooms', and a brush to find fingerprints. But that's only for an old location with minimal facilities. In a haunted house with electricity, his kit expanded to include 'infra-red filters, lamps and ciné films sensitive to infra-red rays, so that I could take photographs in almost complete darkness'.

Price had created, in other words, a portable version of his ghost laboratory.

It is Price's kit rather than Dingwall's, however, that seems to have the lasting legacy, despite the fact that he pinched the idea to have one at all if Dingwall is to be believed. Although surely it would be common sense to take equipment with you when searching a dark house for things that seemingly interact with nature in mysterious ways. After all, Miss R. C. Morton used string and adhesive to set up traps to test the immateriality of the Cheltenham ghost, and William Jackson Crawford squashed putty into tins to gain casts of his 'pseudopods'. Ghost hunters have always used household items and knick-knacks in their pursuit of the paranormal; Price just made his look extra flashy and laid the foundations for ghost-hunting equipment to become a very lucrative business.

Before it was abandoned, Price's laboratory became embroiled in another high-profile case of mediumship. But where Helen Duncan was rather brutally caught out by Price's experiments and equipment, his involvement with the medium Eileen Garrett lauded, rather than challenged, her powers. As the wounds of the First World War began to heal, mediums once more evolved into a new trend of clairvoyance. They returned in part to the style of séances conducted by Florence Cook, claiming to manifest a spirit 'control' or 'guide' who linked them to the afterlife. Rather than materialise the spirit entirely, which had been proven to be an unwise act, these modern mediums would change the tone of their voice and their guide would speak through them. Problematically, and owing perhaps to the rise of cinema which allowed people to experience global images of life and culture in addition to the centuries-long trope of the 'Noble Savage', these guides were often imitations of stereotypical 'wise' Native Americans or ancient Arabian philosophers. For Eileen Garrett, her control was the latter – an Arabian man named Uvani. This wasn't all that separated mid-century mediums from their predecessors, however. Eileen Garrett also led the way for a new trend in predicting the future – something mediums hadn't really tried their hand at before. Indeed, the role of mediums until then had been to discuss what had already passed – the dead coming back to reminisce about their lives – and to describe a spirit's present reality in Summerland. This was especially important in the immediate aftermath of the First World War, as we've already seen, when so many families had no idea what had happened to their sons, brothers and husbands. Mediums were on hand to describe the moment of death from the spirits' point of view. By the late 1920s, however, Europe's relative peace was beginning to fray once again and the future was uncertain. Eileen Garrett and her contemporaries, answering the public's need to know what was in store for them, began to deliver predictions as told to them by their one-dimensional mystic spirit guides.

It all sounds rather doubtful, yet Price found himself drawn in to Garrett's most infamous case. The events began in 1928, while Price was still very much involved in his ghost hunts at Borley Rectory. With the war having necessitated quick, dramatic advancements in aeronautical engineering, pilots who had survived the dogfights over Europe now found themselves involved in ambitious projects to use these new planes for less violent purposes. One goal in particular was to design aeroplanes that could safely and comfortably traverse the Atlantic from Great Britain to the United States. Britain wanted to be the first country to make aeroplanes capable of such a feat – a sort of precursor to the space race the US and Russia would later be involved in. Unfortunately, however, British pilots were continuing to find watery graves as the planes simply couldn't withstand the journey or rapidly changing weather conditions.

One such pilot who died this way was Captain W. G. R. Hinchcliffe, a celebrated veteran whose skills in long-distance, difficult flights were unmatched. But even for him, the journey across the unforgiving Atlantic was too much, and he went missing a few days after he set off in February 1928.

Hinchcliffe would not be laid to rest so easily, however. Eileen Garrett, minding her own business, began to have visions. She was at this point only just breaking out as a medium, and was involved with the British College of Psychic Science under the mentorship of the college's founder, James Hewat McKenzie. One day, while walking through Holland Park in London, she looked up at the sky and saw what historian John G. Fuller described as 'a giant airship emerge' from the clouds – a Zeppelin, or dirigible as they were also known.[6] As she watched it slowly come into view, she was startled to see it suddenly falter in its steady, lazy course across the sky. It shuddered in the air, then, like a sinking ship, turned its nose to the ground as smoke started to billow from it and mingle with the clouds around. The dirigible dropped behind London's tall buildings and vanished from view. Since no one around Eileen was panicking, she knew she must have seen a vision of some sort. At this time, or so she claimed, she was still sceptical of her own abilities – and her immediate thought was that she was hallucinating. She had lived through the Zeppelin raids of the war, so it was understandable that her traumatic experiences were continuing to manifest themselves through strange, hallucinatory phenomena. Even if her vision *was* paranormal, she thought at most that it was an image of the past raids replaying itself to her.

Eileen had only just started to hold sittings in the London Spiritualist Alliance, home to Price's ghost laboratory, and one of her clients was a woman named Mrs Earl. Having lost her son in the war, Mrs Earl was among the bereaved who sought comfort through séances, and a decade later was continuing to actively seek messages from him. At the same time as Eileen was having visions of the doomed dirigible, Mrs Earl was also experiencing something strange, and now sought not to speak to her son through Eileen but to receive her help in explaining this new phenomenon. While at home one evening, Mrs Earl tried her hand at contacting her son through a Ouija board. The planchette began to move in increasingly vigorous circles and Mrs Earl was ready to take down messages from her son. But the first message didn't sound like him; she realised she wasn't in touch with her son at all. She scribbled down the letters: 'CAN YOU HELP A MAN WHO WAS DROWNED'. The messages kept coming, and slowly recognition dawned on Mrs Earl. She had read about Captain Hinchcliffe's missing aeroplane in the newspapers, and the information she was receiving seemed to link with his disappearance. The planchette moved again: 'FOG STORMS WINDS WENT DOWN FROM GREAT HEIGHT'. Mrs Earl was convinced now: Captain Hinchcliffe was dead, had drowned in the Atlantic when his plane disintegrated in a storm, and was desperately getting in touch with her.

Mrs Earl rushed to send a letter about the messages to Hinchcliffe's widow, Emilie, but received no response. She felt herself to be limited in psychic ability, and so returned to her sittings with Eileen Garrett to see if the medium could get in touch with Hinchcliffe in her stead. If she could obtain further information, and perhaps more concrete proof that Hinchcliffe's ghost was trying to communicate, then Emilie would have to take notice. As soon as Eileen went into her trance state, Uvani came forward to tell Mrs Earl that it was indeed Hinchcliffe who had been speaking to her through the Ouija board. Uvani described Hinchcliffe's death: he went off course by 800km (500 miles), and thus didn't have enough petrol to get back on track or indeed back to land, and was defenceless when he hit a violent storm. When Mrs Earl asked specific questions, Uvani was mysteriously vague and didn't know whether Hinchcliffe died at twilight or dawn, nor whether the cause of his troubles was water or fog. Uvani wasn't sure whether Hinchcliffe had one child or two, saying that the captain's spirit was 'very confused'. Perhaps to distract Mrs Earl from her increasingly difficult questions,

Eileen twitched and Uvani, with his generic 'Arabian mystic' accent, seemed to be replaced by a more clipped tone as Hinchcliffe himself briefly spoke through her. He was concerned about his wife, and it seemed more pressing than ever that Emilie should be involved in these sittings.

The president of the London Spiritualist Alliance at this time was none other than Sir Arthur Conan Doyle, and while he had allowed Harry Price to set up his National Laboratory for Psychical Research in the Alliance's building, their friendship was beginning to sour. Doyle didn't approve of Price's scientific equipment, with ultraviolet and infra-red emitters, X-ray machines, state-of-the-art cameras, heat controls and Bunsen burners. Doyle was every bit the traditional Spiritualist, and couldn't stand the thought of mediums being subjected to Price's largely unethical experiments. However, while Price's laboratory exposed many mediums of fraud, he couldn't prove anything against Eileen Garrett. Indeed, he was rather impressed by her, and his approval of her powers was for a little while the only thing that stopped Doyle from cutting ties with Price.

Doyle became increasingly invested in the sittings between Mrs Earl and Eileen Garrett. It was so bizarre, so random, for Captain Hinchcliffe to be communicating with them that there *must* be some higher purpose to it all. He gave them resources and space to continue the sittings, while encouraging Emilie Hinchcliffe to believe in Eileen's genuine powers as a medium. It helped, too, that the notorious Harry Price had given Eileen his seal of approval. Eventually, Emilie became convinced and began to have sittings of her own with Eileen. Despite these positive developments, however, Doyle and Price's friendship finally broke down without possibility of resolution and Doyle kicked him and his laboratory out of the Alliance building.

In the meantime, news began to spread of another attempt for Britain to conquer the journey across the Atlantic. They had momentarily given up on aeroplanes in favour of another means of flight: the dirigible. Eileen was reminded of the terrible vision she saw that day in Holland Park, but her fears were worsened when Hinchcliffe, made stronger by Emilie's presence in the séances, started to finally reveal his reason for reaching across the boundary between life and death in the first place. He was keeping an eye on the dirigible's development, and could see into the future. Like W. T. Stead predicting his own watery grave in the previous decade, Hinchcliffe saw disaster unfolding. The dirigible, called the R-101 airship, would fail and lives would be lost. In a sitting with Emilie and Eileen, Uvani once again withdrew to let Captain Hinchcliffe speak. 'They will start without thinking of disaster,' he told them, 'but the vessel will not stand the strain.' He was particularly concerned because the navigator due to fly in the R-101 was a good friend of his, a man named Johnston. Hinchcliffe had tried to warn Johnston himself, but described him as 'dense' and unreceptive of his messages. Johnston must be warned; it was up to Emilie, Eileen and the well-known names at the London Spiritualist Alliance to stop the R-101 from taking flight.

The airship took a long time to be developed, and it wasn't until the autumn of 1930 that it was finally ready to depart from the hangar. In July of that year, however, Sir Arthur Conan Doyle died at the age of 71. Right up until his death, he was helping Emilie and Eileen to continue communicating with Captain Hinchcliffe, and was passionately lending his weight to correspondence with the R-101's soon-to-be navigator Johnston. Funnily enough, Hinchcliffe was right about the man being 'dense' and unconvinced by Spiritualism; he politely listened on several occasions to Emilie's explanation of the séances and her husband's increasingly desperate warnings, trying his best not to offend the widow, but reassured her that the airship was sound and that there would be no disaster. There was no getting through to him.

Even Uvani, who had taken a backseat in these séances, was becoming increasingly involved in the airship's fate.

Where at first he had claimed that being an ancient Arabian philosopher meant he couldn't understand 'this aviation talk', he'd clearly been doing some research in Summerland's extensive library. A few weeks before takeoff, he was discussing wind resistance, altitude and how their 'gas envelopes attract currents'. Impressive. Even so, Uvani wasn't sure – just as he wasn't sure at the beginning of the case whether Hinchcliffe had one child or two – when exactly the R-101 would meet with disaster. Sometimes he claimed it would be on its maiden flight, and at other séances claimed vaguely that 'she is not going to last'. These were, after all, highly experimental aircraft that were prone to catastrophic failure. In 1923, for instance, France completed work on the Zeppelin airship *Dixmude*, which had been given to its government by Germany as part of its war reparations. It exploded off the coast of Sicily, killing 52 people. Britain had also suffered its own previous failures, too; the R-38 airship disintegrated into the Humber Estuary in 1921, killing 44 of its crew. Prior to its first major journey, the R-101 had undergone several short tests which demonstrated the unreliability of its structure. The fabric covering the 223m (731ft) airship kept tearing, and parts of the structure showed that they were liable to crack and bend. Within this context, Uvani's vague prediction that *at some point* the R-101 would fail becomes decidedly less uncanny.

Still, the gang – both spirits and living – redoubled their efforts to warn Johnston and the crew involved in the R-101's maiden flight. And still, no one would listen. This was the biggest airship in the world, and if Britain could pull it off then it would revolutionise global travel – and make the UK the leading player.

Finally, on 5 October 1930, the R-101 set off from Cardington in Bedfordshire on its first journey across the sea to Karachi, Pakistan. It only made it as far as France before bad weather set in, destroying the structure and sending it tumbling downwards to a forest near Beauvais, where it burst into flames and killed 46 of its 54 crew members onboard.

Meanwhile, Price was being visited by a journalist named Coster who had heard of the sittings going on in his laboratory with Eileen Garrett. But rather than speak to Uvani or Hinchcliffe, he wanted to write an article about Eileen bringing forth the ghost of Sir Arthur Conan Doyle. In the immediate aftermath of the R-101 disaster, they began a new set of séances in which they withheld from Eileen who they really wanted to talk to. Unexpectedly, the first ghost Uvani brought forward was Albert von Schrenck-Notzing, whose sittings with Eva C. we saw in Chapter 4 and who was a friend of Harry Price. Price was thoroughly interested in this, and couldn't understand how Eileen would know of Schrenck-Notzing. This is again confusing; as a medium who had been trained by the British College of Psychic Science and was thoroughly involved with all the key figures of Spiritualism and psychical research, it would be more questionable if Eileen *didn't* know who Albert von Schrenck-Notzing was.

After this mild distraction, Uvani returned to speak through Eileen and resume his new favourite topic: the R-101. He had been speaking with one of the crew members who had met their end on the airship, a man named Irwin. Irwin had told Uvani in astonishing detail exactly what went wrong. The engines were too heavy to keep the ship comfortably airborne. Eileen talked quickly in her trance; almost too quickly for Price's secretary Ethel Beenham to scribble down her shorthand notes. Over and over, Irwin repeated this information, but also listed an immense number of issues, from the oil pipe being clogged to the fuel injection and air pump failing. The technical language was that of an expert and rolled off Eileen's tongue. When Price quizzed her about certain details, she answered immediately and with fluency about the inner workings of the R-101 airship.

Crucially, however, Irwin told those at the sitting that it was pressure and heat, combined with the ship drifting into an electrical storm, that caused it to explode. But it didn't explode; it caught fire when it hit the ground. And it wasn't an electrical storm that caused the damage to the structure, but the violent winds. Before any further questions could be asked, Irwin drifted away and was replaced with Sir Arthur Conan Doyle (who rather rigidly introduced himself with, 'Here I am. Arthur Conan Doyle.'). The R-101 disaster became all but forgotten as Price and Coster engaged the late writer in conversation through Eileen.

While I was in the British Library researching Eileen Garrett, I was reading her autobiography. On one page, scrawled in the margins, was some angry graffiti about accusations of her mediumistic talents being fraudulent, referencing the R-101 case as Exhibit A. It took me a while to figure out what was meant by this. I'll admit that I was initially impressed by the technical details and the overall drama of the case. But Uvani's knowledge of technical details is impressive because I, like Price and Conan Doyle and Emilie and Mrs Earl, haven't any particular understanding of aeronautical engineering. Looking at it that way, you can see that Eileen was going for a scattergun approach – give a huge range of information, and hope that some of it will be correct. She 'predicted' the R-101's downfall because the vast majority of Zeppelin and dirigible projects had ended in fatal disaster. To read enough about past catastrophes would be to identify common faults and errors. When the R-101 did fail, it would likely be down to the same causes as previous failures. For Garrett, this was inevitability dressed up as mystical prediction.

It's peculiar that Price would be so invested in Eileen Garrett when he was only ever sceptical about Helen Duncan. But perhaps, here, we see more of the side of him that turned the haunting of Borley Rectory into a fraudulent publicity stunt. The R-101 disaster was major news – wouldn't it skyrocket his fame to demonstrate that a medium in his ghost laboratory had predicted the entire thing? Nevertheless, it didn't bring him the notoriety he was searching for, and Eileen's predictions and séances are barely mentioned in the history of the R-101 airship.

While Price's ghost laboratory was short-lived and never completely replaced, laboratories dedicated to psychology would soon begin to investigate ghosts and uncanny phenomena under a new branch of the discipline: parapsychology. In particular, parapsychologists in the mid-twentieth century began to wonder whether some spooky occurrences were caused by a telekinetic power within humans. There was one type of ghost that seemed to align with this theory: the poltergeist. Noisy, violent and persistent, the poltergeist has tormented households for hundreds of years. But after the Second World War, once people began to abandon the séance, the poltergeist seemed to be the ghost hunter's most frequent nemesis.

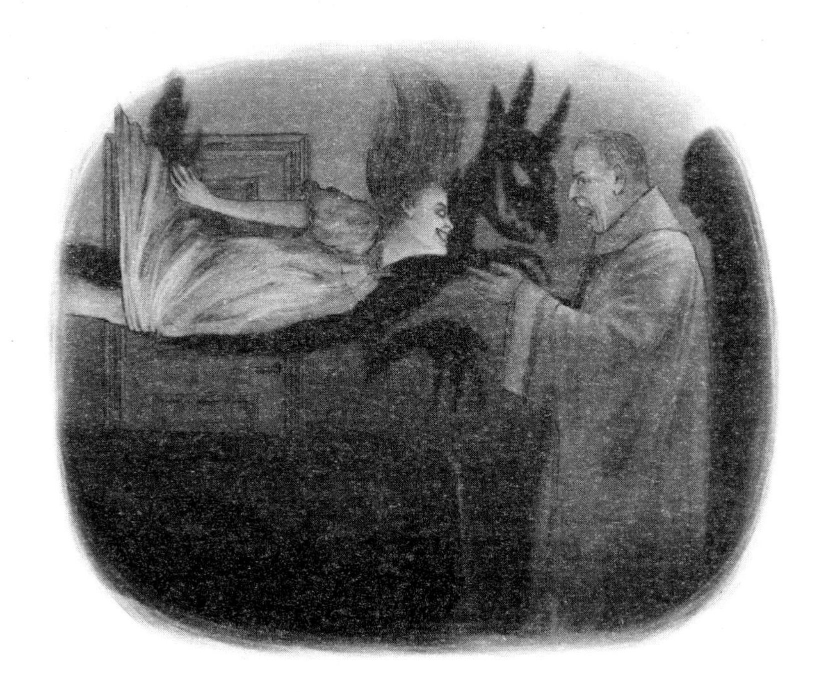

CHAPTER EIGHT

The Poltergeist Next Door

In November 1960, in a rural Scottish village called Sauchie, 11-year-old Virginia Campbell began to hear a strange noise whenever she went to bed. It had been a time of massive upheaval for the child; her parents were undergoing some financial difficulty and so had to accept jobs far from their home town. Rather than take Virginia with them, they had installed her in the care of her aunt in Sauchie, where she would remain and go to school until her parents were in a better position to look after her again. It was a strange new house in a strange new location; she would have to start a new school, make new friends and adjust to life without her parents.

And now, this. It began as a dull, repetitive sound, 'like a bouncing ball', in the bedroom, but there seemed to be no visible cause.[1] Virginia slept in the same room as her cousin, Margaret, and they found the noise followed them one night when they went downstairs to tell Margaret's parents. The noises continued, and other strange phenomena began to occur. A few days later, Virginia was sitting quietly in the living room. The sideboard next to her suddenly sprang forwards a centimetre or two, then moved back again on its own accord. Then a solid, sturdy chest of drawers full of linen rocked and seemed to glide across the floor – a full 45cm (18in) – before coming to rest again. The little house in Sauchie seemed alive with an unseen extra guest, one who banged and thumped the walls and seemed especially adept at moving hefty pieces of furniture.

Whatever was causing the strange events was stalking Virginia. At school, her teacher, Miss Stewart, watched in horror as an unoccupied desk behind Virginia rose a centimetre off the floor and then came back down again, slightly askew from its original position.

Local news caught wind of the sensational phenomena that surrounded this little newcomer to Sauchie. It was dubbed a 'poltergeist', a German term meaning 'noisy ghost', and, in the second half of the twentieth century, it was the sort of quirky story that sent journalists wild.

Trevor C. Hall, who had a hand in the controversial take-down of Harry Price, *The Haunting of Borley Rectory*, read the newspaper articles with profound interest. He sent some clippings to his friend, Dr A. R. G. Owen, who was one of the overseers of the 1970s Philip Experiment in Toronto. All the characters in the history of ghost hunting seem to become linked in one way or another. Owen's interest was piqued; he had been working with the Toronto Society for Psychical Research on poltergeist phenomena, which seemed not only to be the fashionable research topic among ghost hunters, but was also becoming increasingly reported both in the media and to psychical research societies.

Owen travelled to Sauchie to make his own inquiries, and soon became embroiled in the battle between Virginia and the poltergeist that trailed her like a shadow. He was not so much interested in the poltergeist itself, however; his attention was focused on *Virginia*.

Poltergeists have been recorded for millennia, and are characterised by loud noises, movement and levitation of objects by unseen hands, and the throwing of stones and other small items. One of the first ghost stories involved poltergeist phenomena, and dates back to ancient Rome; stones were said to be thrown inside a house, and the property was condemned. In 1575, in a rented house in Tours, France, the new tenant, M. Gilles Bolacre, soon found himself being terrorised by a noisy ghost. He appealed to the local tribunal that the house was not fit for human habitation and that his contract with the landowner should be revoked, which was granted by the court.[2] But over the years, and especially with the development of ghost hunting and psychical research, poltergeists have been understood as something quite separate from hauntings. As we've already seen, haunted houses and bog-standard ghosts have a lingering, intangible quality to them; they repeat their actions over and over, sometimes as blissfully unaware of the living as sleepwalkers. They are said to move as though in a dream state, and some researchers even consider them to be echoes replayed by the very fabric of buildings themselves. Ghosts may be 'living' versions of the dead, but they are still very much dead. Poltergeists are a different story altogether. Poltergeists are wholly *alive*. They are tempestuous, they respond in real time to events or statements, seem to latch on to and obsess over one particular person within the house, and, moreover, only seem to get stronger whereas the ghost only ever seems to fade with the passing of time.

Stone throwing, or lithobolia, is a particular form of poltergeist activity that seemed particularly rife during the height of European imperialism in the late nineteenth and early twentieth centuries. There are numerous cases of wealthy white families, living and making their fortunes in colonised countries, who were terrified out of their homes by supernatural forces. In 1937, in Johannesburg, the Ricks family home began to be pelted with stones – around 80 in total fell on the roof and in the courtyard, and, interestingly, some of the stones weren't geologically local. The Ricks family had the police watch over them, and when they tried to seek shelter elsewhere they found that the poltergeist followed them; the phenomena only ceased when they hid in a hotel. They blamed the children's nanny, an 11-year-old girl of Creole descent, and dismissed her.

Similarly, in 1906, the Society for Psychical Research printed a report in their *Journal* sent to them by an associate member named Mr Grottendieck from the Netherlands. Grottendieck had been in Sumatra working for a Dutch oil company, and was spending some time off in the jungle accompanied by 50 local people 'for exploring purposes'.[3] One night, in the camp that consisted of wooden huts covered with large leaves, Grottendieck settled down under his mosquito net. He was suddenly awoken at one o'clock in the morning by something falling down and hitting the ground near his head. Jolting in fright and hurriedly lighting his kerosene lamp, he found a small black pebble. More of these pebbles started to rain from the roof in a 'parabolic line'. It tells us quite a lot about Mr Grottendieck's personality that he rudely woke up an Indonesian boy sleeping in the next room of his hut and sent him outside, in the dead of night, to 'examine the jungle up to a certain distance'. The boy came back empty-handed, and Grottendieck told him to investigate the rest of the hut while he stayed put.

The stones continued to fall, and when Grottendieck tried to catch one, he found that they changed direction in the air, skirting around his hand. Moreover, while they appeared to fall through the roof, there were no gaps or holes in the leafy canopy.

Grottendieck was getting wound up, and started to suspect the rest of his exploration party were playing a practical joke on him. Insecure at the thought of being the butt of the locals' joke, he grabbed his rifle and fired five shots at random into the jungle. This obviously startled the boy quite badly, who said the stones were the fault of Satan and ran off, never to be seen by Grottendieck again. Grottendieck interpreted this as the boy being so scared of the poltergeist that he fled; more likely he was terrified of being shot by this raging madman.

As soon as the boy fled for his life, the stones stopped falling. But they littered the ground of Grottendieck's hut and, picking one up, he found that they were warm to the touch. They didn't disappear by the morning, meaning he didn't hallucinate them, and he counted around 20 stones in total. In case Grottendieck's act of waking up the boy, sending him alone into the dark jungle and then firing a gun madly towards him didn't tell you everything you needed to know about him, he ends the first part of his report with a particularly flat attempt at comedy: 'The worst part of this strange fact was that my boy was gone, so that I had to take care of my breakfast myself, and did not get a cup of coffee nor toast!'

The SPR, intrigued (or perhaps bemused), wrote back to Grottendieck to glean further information from him that would help them to draw conclusions about the phenomena. Grottendieck firstly noted that these instances of stone-throwing poltergeists are common in Dutch colonies, and recalled reading about various cases reported frequently in newspapers. In answer to the SPR's questions, Grottendieck revealed some interesting aspects of the phenomena – it's odd that he neglected to mention them in his initial report. He insisted that the boy was not the culprit; the stones fell while the boy was in the other room, sleeping, and continued to fall while Grottendieck sent him into the jungle on his own. Pressed on the temperature of the stones, he went into further detail and explained that they were hot – certainly hotter than they should have been if they had simply been held in the hand or in a pocket – and likened them to meteorites. Weirdly, too, he described the way the stones fell slowly, abnormally slowly, as though feather-light, yet they touched the ground with a loud thud as though they had slammed down with violent force.

The SPR, in their criticism of the case, still thought it was likely the boy was causing the phenomena. It certainly explains why he bolted as soon as Grottendieck began firing at the poltergeist – presumably there was something of a language barrier between them, and perhaps the boy thought he would be Grottendieck's next target when he found out that he was causing the stones to fall. Moreover, three years had passed between the incident and Grottendieck's report to the SPR. While the idea of the stones feeling hot and seeming to fall down slowly are peculiar details in the case, it may well be that Grottendieck had misremembered and exaggerated certain aspects. The stones were warm to the touch, but in Grottendieck's fear he may have thought they felt hotter, and remembered it that way; the stones perhaps fell at a normal pace, but having been jolted awake and feeling threatened enough to fire his gun into the dark, Grottendieck may have felt that the moment had slowed down.

What both the Johannesburg and Sumatra cases demonstrate is that poltergeists are taken more as a personal threat than stereotypical ghosts. It's interesting that both cases, representing hundreds more like them, demonstrate underlying tensions between coloniser and colonised, and these tensions bubble up to the surface in the form of violent phenomena. In these cases particularly, they show, I think, a deep-rooted insecurity and guilt within the minds of imperialists who feel they have incurred the wrath of paranormal forces within a culture they don't understand. There's always a sense of being targeted, of being hunted, and of being made to feel vulnerable and powerless in the face of an enemy they cannot control. This, ultimately, is the poltergeist: something faceless, persistent, malevolent and uncontrollable.

There is one trend in poltergeist cases that defines them; their 'target' is often (although, as we'll shortly see, not always) a child on the cusp of puberty. In the latter half of the twentieth century, as poltergeist cases seemed to be springing up more regularly, psychical researchers and parapsychologists began to explore the phenomena in relation to their research on psychokinesis, or PK as they often called it. They had a theory that poltergeists *weren't* ghosts at all – that they weren't the trapped souls of people who had been previously alive – they were, in fact, a fractured piece of a pubescent child's mind that was manifesting as some sort of telekinetic energy. The 1970s, in particular, saw a flurry of research among parapsychologists investigating children's potential to exert a mysterious force. This was the age of Uri Geller and spoon bending, and British psychical researchers and physicists were getting children to try to bend and manipulate metal that was sealed in glass domes. Among the most influential scientists was Professor John Hasted, a physicist working at the University of London. His son, also John Hasted, got in touch with me and, together with my similarly ghost-obsessed colleague Dr Luke Thurston, we met him when he was passing through West Wales. In his bag, he had a number of bent pieces of metal from his father's collection, including a spoon with such a perfect, tiny, uncannily intricate twist at the top of the handle that I confess I experienced a little shiver. John told us about his father's work, and showed us photographs of glass baubles with explosions of jagged metal inside, like sea urchins, and stated that these were paperclips that children had manipulated in experiments. His father, Professor Hasted, thought that these abilities were the cause of poltergeist phenomena (which, ironically, he was experiencing at home – although the target seemed to be his wife rather than his son). As anomalistic psychologist Professor

Chris French illustrates, the child was seen by parapsychologists as the 'focus' whose 'inner psychological turmoil somehow becomes externalised in the form of psychokinetic energy.'[4] The child seemed unaware of this part of themselves, and didn't know that the poltergeist who terrorised them so acutely was actually a fragment of their own personality.

Equally, from the sceptical point of view, the child could also be very good at playing elaborate pranks.

Dr A. R. G. Owen had a good feeling about the Sauchie poltergeist, though. For one thing, Virginia was now being regularly visited and observed by a number of people. Miss Stewart, her teacher, had become more involved, and was keeping reports of the strange phenomena that were happening at school. The local reverend, Mr Lund, was often called to investigate the phenomena from a religious perspective, while three doctors, Dr Nisbet, Dr William Logan and his particularly sceptical wife, Dr Sheila Logan, had been keeping meticulous reports for Owen about what had been happening to Virginia.

Together, the group turned into an unlikely gang of ghost hunters, observing and noting the poltergeist as objectively as they could. On Saturday 26 November, Dr William Logan saw how the pillow beneath Virginia's head rotated, despite her hands being nowhere near it. The bedclothes, too, puckered in a strange way that was impossible to replicate. The following Monday, Miss Stewart noted how Virginia came up to her desk to ask for help with a worksheet. As the girl stood, hands clasped behind her back, a little way away, and Miss Stewart watched as the blackboard pointer on her desk began to vibrate and move until it fell to the floor with a clatter. In terror, Miss Stewart put her hand on top of the desk and felt it trembling; it skittered across the floor until it had almost spun entirely around. Virginia stared, unmoving, a metre or two distant from the desk, while all this occurred.

Virginia then moved away from Sauchie, and lived with her mother in Dollar where she worked. The Logans visited her in the new house, and found that the poltergeist had followed the girl there, too. They heard several instances of disembodied, but intelligent, knocking and thumping noises. Sheila, being previously extremely sceptical of the whole affair, left the house a little subdued, having no explanation at all to offer as to how Virginia could be making the sounds herself.

After a while, once Virginia had settled back in with her mother away from Sauchie however, the phenomena began to fade. Owen's group of investigators visited occasionally, but remarked how much happier Virginia was now, and that the poltergeist had mostly vanished. If we subscribe to the idea popularised by parapsychologists that poltergeists aren't mischievous souls of the dead but a kind of telekinetic personification of childhood distress, then the fact that Virginia's paranormal pest left her alone as soon as she was settled back with her mother certainly seems to attest to this theory. Indeed, many poltergeist cases dwindle when there's a change in circumstances for the focus of the entity's violence; the stressful period has passed, or the living situation improves. During my ASSAP investigator training, instructor Steve Parsons told us how poltergeists are intense, violent, but also short-lived. The phenomena gets stronger and stronger, but often disappears just as suddenly as it arrived, and where stereotypical ghosts can linger for hundreds of years, poltergeist cases rarely last more than a few months (unless there's money to be made, as we'll see in a later chapter).

In the Senate House archives, there is a collection of notes – and the original reports from Lund, Stewart and the Logans – that Dr Owen used to write his book *Can We Explain the Poltergeist?* (1964). Amid the newspaper clippings is a piece of paper with several small photographs pasted to them. One of them depicts the Sauchie poltergeist house.[5] The fact that the photograph is black and white doesn't help matters, but I was struck by how drab the building is. It is a basic, square semi-detached house with smooth concrete walls that seem to retain the damp from the Scottish rain. A small patch of grass extends at the front of the house, leading to rough concrete posts holding up a messy, unwelcoming wire fence. Creepily, two almost ghostly figures can be seen staring out at the photographer from the living room window. While there's nothing to suggest that the atmosphere inside of the house was anything but warm and friendly, and none of the reports describe any kind of hostile environment between Virginia and her aunt, there's no denying that the house looks deeply depressing from the outside. Virginia, in a state of upheaval and facing the daunting prospect of a new school, may well have acutely experienced this. After all, many of the poltergeist incidents did occur in the classroom; while Virginia's teacher seemed like a sympathetic and kind woman, living in a new home with a relative and attending a new school is upsetting for any child, regardless of how loved they are.

The Sauchie case is one of a larger trend of poltergeist cases in the second half of the twentieth century. We saw how after the First World War, séances came back into fashion as a way of coping with the aftermath. After the Second World War, while some mediums rose to prominence for continuing this tradition, the focus shifted; ghosts weren't being summoned into living rooms any more, they were cropping up of their own accord – and they were wreaking havoc. This may well have been in answer to the anxiety of enemy bombs. Families couldn't control the violence raining down on them from above, so perhaps creating poltergeists to destroy their houses from the inside gave them some subconscious feeling of authority over the matter. But the majority of high-profile poltergeist cases came a few decades later, particularly in the 1960s, 70s and 80s, when children didn't live with the anxiety of being bombed in their own homes.

The British poltergeist of this era seems to spring from another kind of social issue, but one still very much related to our connection to the idea of home: council house provision. This began in 1900 as a way to provide adequate accommodation for working-class people in the UK, and to reform slums and generally lift the overall health and well-being of the public. In the 1940s and 1950s, hundreds of thousands of prefabricated homes, cheaply built and meant only to last 10 years, were constructed as accommodation for the working class. Subsequently, whole villages and towns were built around the council house system. But by the 1970s, when Prime Minister Margaret Thatcher's Conservative government saw to drastic cutbacks of housing provision and maintenance, public opinion of council houses changed dramatically. The reinforced concrete of many council houses was damp, many were too small, and now they were becoming underfunded and neglected. Moreover, council house estates were developing a reputation for being 'rough' areas with high crime rates and problematic neighbours.

For children, particularly those of the age of puberty who were beginning to understand more about the world and their place within it, living in a council house could be particularly difficult. As Andrew Green notes in his *Ghost Hunting: A Practical Guide* (2016), one of the first things to identify when investigating a poltergeist case with a child 'focus' is how long ago they moved into the property, especially if the property is a council house. 'Little calming influence', he says, 'is received from the parents at this time, as they too are undergoing the upheaval of the routines and behaviour patterns' and will often shoo the child away for being a nuisance.[6] After all, when it involved a council house, the family had likely moved there for an unpleasant reason, whether it was because of divorce or a difficult financial situation. If we subscribe to the idea that poltergeists are a product of a child's anxiety – regardless of whether we believe the phenomena is faked or part of some psychokinetic ability – then it's understandable that living in a sub-standard council house would provoke such occurrences. Green notes that the poltergeist was simply a way for the child to 'reassert' their presence in the household.

One of the most high-profile cases seems to align with this idea. Between 1977 and 1978, during the peak of Thatcher's neglect of council house provision and maintenance, a family in Enfield were terrorised by a violent poltergeist who fixated on 11-year-old Janet Hodgson.

The poltergeist truly made itself known in the evening of 31 August 1977. In the days leading up to the first explosive night, Janet and her brother, Johnny, had been complaining of noises, of their bed shaking, but nothing had become drastic enough for Mrs Hodgson to take notice. On 31 August, however, Janet and Johnny complained of 'shuffling noises'.[7] Rather than haughtily dismiss her children again, Mrs Hodgson listened. And she heard it; as eyewitness Guy Lyon Playfair describes, she 'thought it sounded like somebody moving across the floor in slippers, and she was sure that whoever or whatever was making the noise, it could not be either of the children.' Things soon took a dramatic turn, with loud bangs coming from the wall and heavy pieces of furniture sliding across the floor, pushed by unseen hands and witnessed by Mrs Hodgson and her children.

Unsure of what to do, Mrs Hodgson called the police. It seems like a silly thing to do, but she was frightened, confused and trying to keep her gaggle of children calm. While writing this book, I holidayed for a week in a cold, mouse-ridden fifteenth-century farmhouse with my mum, my sister, her husband and his family. Walking into the large kitchen, thinking my mum and I were the first ones back after we'd all had our respective days out, we came to find all the drawers pulled open, all the cabinet doors wide on their latches, and the heavy wooden chairs upside-down on the table. We both gasped, and, not renowned for being tactful in such a situation (and forgetting Rule #1 of my ASSAP training), I said, 'Mum … this is classic poltergeist activity.' We both were, in no uncertain terms, freaking out. I remember feeling a strange thrill that finally, *finally*, I was experiencing something inexplicable; but I also felt a very curious vulnerability. For a moment, we were both stunned, a little panicked; we simply didn't know what to do. It turned out, of course, to be a very well-done prank courtesy of my brother-in-law's brother, but it made me appreciate a little more just how such uncanny phenomena can drive a person to cling to any possible glimmer of safety and protection.

At the Enfield house, Constable Carolyn Heeps and her colleague soon arrived and found Mrs Hodgson and her children being comforted by their ne ighbours, the Nottinghams, who, it must be said, are painted as incredibly supportive in Playfair's description of the case. But if the two officers were exchanging suspicious and sceptical glances, their opinions were soon drastically changed when Johnny pointed to an armchair in the living room. It began to shudder, then moved quickly towards the kitchen, with Carolyn Heeps reporting a total distance covered of 'three or four feet'.

Over the next few days, the phenomena began to get more violent, more targeted. Small projectiles started to be launched at people – Lego bricks, marbles. Playfair notes that they 'would just zoom out of thin air and bounce off the walls, or drop straight to the floor as if they had come through the ceiling'. It seemed relentless, and was becoming increasingly dangerous. The police tried their best to calm the Hodgsons, but they were at a loss as to any practical help they could give. Peggy Nottingham took action, and contacted the *Daily Mirror* newspaper to see if they could run a story on their plight, in the hopes that it would attract the attention of someone who might be better equipped to help them. As we saw with the Sauchie case and Eric Dingwall's scrapbooks of poltergeist report clippings, the newspapers at the time lapped this up. They still do; not a week goes by without my dad emailing me an article from a local Welsh news website about ghosts, poltergeists and haunted dolls. Interestingly, Playfair describes that the *Daily Mirror*, in a rare departure from the rest of the UK tabloids, had a rule among their reporters that they wouldn't run sensational stories of paranormal activity because of the way Harry Price tarnished their reputation with his fantastical (and often fraudulent) stories.

There was, however, something about the Enfield story that led the *Daily Mirror* reporters to break their own rule and come to investigate. After all, with witnesses from two different households and the police, as well as the regularity and violence of phenomena, this was no ordinary ghost story. The involvement of the *Daily Mirror* led to the outcome Mrs Nottingham hoped for, and the Society for Psychical Research made themselves known to the Hodgsons. Maurice Grosse, a moustachioed, kind-faced investigator with the SPR attended to see whether the case had any merit or was, in fact, just another melodramatic tabloid story. But any initial doubts he had on the journey to the Enfield house were quashed the moment he stepped through the door and noted the genuine fear on the faces of those within.

He took the case seriously from the outset, and encouraged Mrs Hodgson to stop being a passive victim in the situation. He taught her a few basic principles of investigation, advising her to keep fastidious notes of the phenomena, and it helped to soothe her nerves. Indeed, even the poltergeist seemed to settle down in the presence of Maurice Grosse and his calm, methodical approach, with activity easing in the days following his arrival.

It was only a temporary respite. A few days later, the phenomena struck up again with renewed and violent intensity. Grosse, accompanied now by three *Daily Mirror* reporters who were now deeply invested in the ongoing story, were holding an overnight vigil on the upstairs landing. From Janet's room, there was a sudden crash, and the men burst in to find a chair in her bedroom had moved several metres from its usual position and tipped over. Janet woke up, crying. Crucially, though, no one saw it actually happen.

Grosse was perplexed, wanting to pin the phenomena on Janet but finding too many holes in his theory. Phenomena such as the upturned chair could easily have been done by her and, indeed, much of the activity seemed to happen around her. In parapsychological terms, she was undoubtedly the 'focus'. But, for every nine instances of phenomena that occurred when Janet was nearby, there was a tenth that happened well out of the way. As well as Johnny, Janet had two other siblings, Margaret and Billy – but even so, spontaneous phenomena occurred even without any of the children near.

Other news and media outlets picked up on the *Daily Mirror*'s particular interest in the case, now legitimised by a real ghost hunter in Grosse. LBC Radio's Mike Gardiner became involved, finding his way to the Enfield address through clues and luck despite the *Daily Mirror* taking pains to keep the information classified. He interviewed the Hodgsons and Grosse, to an estimated quarter of a million listeners. The British public were now hooked, and the Hodgsons and Grosse now found themselves at the centre of attention.

And poltergeists love attention.

It was at this point in the case that Guy Lyon Playfair, a well-known parapsychologist, entered the fray. His description of Janet immediately conveys his suspicion of her – he calls her 'all energy' and accuses her of having 'an impish look', sympathising with those who came to suspect her of producing the phenomena through fraudulent means. Despite the fact that he never outwardly points the finger at Janet in his 1981 book *This House is Haunted*, and, in fact, pushes the idea that the Enfield poltergeist was a genuine haunting, his account seems to take on a more dubious view of events once he starts to describe the things he witnessed firsthand.

With an extra person now crammed into the house, the poltergeist activity only seemed to get stronger. Hefty pieces of furniture such as a large armchair were being upturned and, now, puddles of water with no obvious cause began appearing in the kitchen. Like Virginia in Sauchie, and, perhaps a crucial aspect of the case, Janet's schoolwork began to suffer. Just as things in Virginia's classroom had moved

– much to the horror of Miss Stewart – Janet began to complain that her chair had started to shudder and twitch underneath her, and her exercise books were messy and full of jagged scribbles. It's interesting how much Dr Owen and the people surrounding Virginia, including Miss Stewart, seemed to pay close attention to Virginia's wider situation and how she was faring at school. There's not much to suggest that this ever happened with Janet, and I wonder if doing so might have shed a little more light on the origins of the poltergeist.

Still, the attention surrounding the council house in Enfield continued to build. It was being taken seriously enough that a production company from the BBC arrived to create a documentary feature, recording as much evidence as they could.

In the presence of so many watchful eyes, the poltergeist evolved again. Playfair started to try to communicate with the entity through Spiritualist methods such as automatic writing, and in doing so coaxed out a rather terrifying personality. Only trusting Mrs Hodgson with the task of automatic writing so as not to scare the children, he left notepads and pencils around the house, inviting the poltergeist to leave messages. While no messages initially appeared on the notepads, they did find a note on a scrap piece of paper on top of the fridge that read: I WILL STAY IN THIS HOUSE. DO NOT READ THIS TO ANYONE ELSE OR I WILL RETALIATE. Later, they found a torn page in Janet's exercise book that matched this scrap of paper.

This was the beginning of a sinister new development: the poltergeist, through Janet, began to speak. Playfair calls it 'the Voice', but it soon develops its own personality as a cantankerous man named Bill. Recordings were made of interviews with 'Bill', and can still be found on YouTube. The grainy background of the tape only serves to make the whole thing feel much more sinister; in response to clipped 1970s accents, Bill's gruff, simple speech is effective at raising the hairs on the back of your neck. He claimed he had died in the armchair that frequently tipped over and moved in the living room. But some of the things Bill came out with are, to put it mildly, things you wouldn't expect a chain-smoking old man to say. In a somewhat awkward part of the case, Playfair goes into detail about how Janet's first ever period began during the investigation. During a chat with the poltergeist, it said: 'WHY DO GIRLS HAVE PERIODS?' Despite Playfair trying to steer the conversation on to other topics, the Voice wouldn't let it go, and kept demanding to have menstruation explained to it. This, for Playfair, 'was a bit too much' and only made him more suspicious of Janet's culpability in the whole affair. After all, if the ghost of an old, grumpy man managed to pierce through the veil and communicate with the living, why would its conversational topic of choice be the ins and outs of menstruation? Moreover, while the voice is deep and eerie, it's not impossible that Janet could be manipulating her vocal cords to sound like that. One of my own party pieces at that age (or so I thought) was impersonations, and I knew a boy at school who could make a sound uncannily like a dentist's drill; children make funny noises. I think because we become used to the flat tones and seriousness of the adults around us, we forget just how much children enjoy, and are good at, changing their voices to sound strange or silly.

The presence of 'Bill' seemed to tip things over the edge. It was too much for some investigators, and, as we've just seen, Playfair's own doubts became more pronounced. From then on, it all began to fall apart a little. Since the poltergeist was now displaying intelligence and sentience through 'Bill', investigators asked him to perform paranormal tasks in front of them to validate himself. But as long as people were directly watching Janet's every move, nothing happened. It was only when they left Janet alone did phenomena occur, but even then there was evidence that Janet had simply done it herself – a slipper that Bill had been asked to dematerialise turned up under her mattress. Playfair's tape recorder, too, disappeared; when he eventually found it, still running, it had recorded the unmistakeable sound of Janet picking it up and moving it. To make matters worse, Playfair described how 'she seemed to be enjoying every minute of it all'.

Nearly 50 years after the fact, the debate about the Enfield poltergeist still rages. Photographs of Janet and Margaret in their bedroom were taken remotely from a room downstairs by Graham Morris, and are now infamous: one depicts Janet apparently levitating, lifted off her bed while screaming in terror. Of course, she could simply be in mid-jump. The Enfield house was also briefly visited by American demonologists Ed and Lorraine Warren, whom we'll unfortunately meet properly in the next chapter. Now the protagonists of *The Conjuring* film franchise, *The Conjuring 2* (2016) dramatises their involvement in the case, making it seem as though they were the major investigators. While the film includes moments of doubt and suspicion surrounding Janet's character, it ultimately, and very dramatically, presents the case as being genuine – with the haunting being caused by a demonic nun (of course). For every sceptical re-examination of the evidence, there's a new dramatisation that frames the Enfield case as a true haunting.

The thing that bothers me about poltergeist cases involving the stereotypical child 'focus' is how much a sceptic relies on a child's intelligence to be so very cunning and manipulative. At age 11, I was described in school reports in similar ways to Virginia and Janet – bright, quiet, a 'wry' sense of humour – but I could barely plait a friendship bracelet and wouldn't have been able to manipulate my way out of a paper bag. I'm not saying the parapsychologists have it right, that there's some sort of strange psychokinetic force at work here, but there's something about poltergeist cases that makes me uneasy. It's a lot of grown-ups bearing down on a child, just waiting for them to slip up. It's not about ghost hunting in the traditional sense of holding vigil and observing for evidence of the intangible; it's about doubt and suspicion, about catching culprits in the act. Psychical researcher Sacheverell Sitwell remarked that it often 'takes more than one person to make a poltergeist', and in many cases the whole family is brought under scrutiny, as with the Enfield haunting.[8] Although the adults surrounding Virginia in Sauchie seem somehow kinder and more supportive than those involved at Enfield, there's always a distinct lack of ethical considerations, a distinct lack of interest in what might *drive* a child to fake the phenomena.

This was not, however, the case with the Thornton Heath poltergeist. In 1938, parapsychologist and psychoanalyst Nandor Fodor, member of the International Institute for Psychical Research, was alerted to poltergeist activity that surrounded an adult woman. Fodor approached ghost hunting from his expertise in psychology and psychoanalysis, developing the idea that poltergeists, in particular, were manifestations of the troubled minds of the living rather than phenomena caused by the spirits of the dead. He had experience with poltergeists in the past, and was one of the lead investigators – along with, of course, Harry Price – on the unusual and notorious Gef the Talking Mongoose case on the Isle of Man, in which the troublesome entity declared himself to be a mongoose who left ectoplasmic pawprints and ceaselessly chattered and scrabbled around the house. Fodor stayed at Gef's house for a week without experiencing a single incident, and chalked up the sensational phenomena to the homeowner, James Irving, while others pinned the poltergeist on James' 13-year-old daughter Voirrey. This analysis was typical of Fodor's interpretation of poltergeist phenomena; in general he believed a poltergeist was 'psychosomatic dissociation, a mental split' that could, somehow, cause the telekinetic activity so often recorded in such cases.[9] When invited to investigate a case, he always looked to the flesh-and-blood residents for answers; the poltergeist was largely a matter for psychoanalysis.

This was the theory on which Fodor approached the Thornton Heath case. At the centre of it was Mrs Forbes, an attractive housewife who, under the surface, was incredibly troubled. The phenomena, at first, mostly included the violent levitation and throwing of objects, which Fodor almost immediately began to witness once he started to investigate the case. There was almost no limit to the activity; within the space of just two hours, Fodor records that a clock, a chair and a glass were flung through the house – although, having never seen the object *begin* its journey, and with Mrs Forbes always being elsewhere, Fodor saw these events as non-evidential.[10] However, there were still some bizarre occurrences. Mrs Forbes held a glass tumbler, and it suddenly, without her moving her hand, flew with great speed at the window and explosively smashed – Fodor marked this in his notes as '*impressive*'.

There's a sense throughout his discussion of the case that Fodor was more interested in Mrs Forbes herself than the phenomena. He wrote how terrified she was after each smashed glass or broken ornament, and how her heart beat at an alarming rate when there was a flurry of violent paranormal activity. It was due to these reactions that Fodor was convinced that Mrs Forbes was genuinely frightened and shocked by the phenomena, and thus couldn't be faking it. But we've seen that before, when Guy Lyon Playfair described Maurice Grosse's first impression of the Enfield house and its atmosphere of dread. In the Enfield case, many details were suspicious; in Fodor's Thornton Heath case, he would also soon learn that just because the poltergeist's 'focus' is afraid, it doesn't mean they are completely blameless.

Nevertheless, Fodor did retain a healthy amount of scepticism, even when faced with phenomena he couldn't explain. It wasn't necessarily the phenomena itself that interested him; while he was a parapsychologist and psychical researcher, it was with his psychoanalyst hat on that he approached the Thornton Heath case. The more he got to know Mrs Forbes, the more he saw that she was an incredibly troubled woman with a multitude of past and ongoing health issues and some sort of buried trauma that he believed was the root cause of the phenomena. When he began to catch Mrs Forbes in the act of faking phenomena, there's a sense of disappointment that, yet again, definitive proof of the paranormal eluded him. Yet there's also a feeling of sympathy, of pity, that he was faced with a woman so churned up inside that she was resorting to these elaborate performances. Claiming that her wedding ring had gone missing, Mrs Forbes was seen by Fodor to put her hand in her pocket; stopping her from doing so, he found the wedding ring in there – she was clearly about to set it down somewhere and declare that it had mysteriously returned in a random place in the house. Even more curiously, the Forbes' lodger, Mr Simmons, one evening distracted Fodor's attention so that Mrs Forbes had chance to make 'reappear' Fodor's missing magnifying glass which she had hidden down her blouse. Fodor, to his credit, didn't jump to the conclusion that Simmons distracted him on purpose – it could simply have been that Mrs Forbes was waiting for her opportunity.

The more Fodor got to know Mrs Forbes, the more tangled up he became in a long and dangerous series of mind games and manipulation that almost cost him his career. One of his biggest mistakes was to try to separate Mrs Forbes from the poltergeist by conducting a series of 'controlled' experiments with her in a room away from the tumultuous environment of her home. He encouraged her, with no small sum as payment, to hold séances in the International Institute for Psychical Research and, like the sensational mediums of the Victorian era, she began to demonstrate materialisation abilities. Indeed, after a little while, the paranormal phenomena surrounding Mrs Forbes became so far removed from the poltergeist activity within her home that it was possible to forget how Fodor's involvement began in the first place.

During these séances, in which so many unfamiliar people sat around, fussed over and observed Mrs Forbes, the phenomena became less like a chaotic poltergeist case and more like the materialisation mediums of the previous century. She thrived in the attention being paid to her. Miraculous objects started to appear in the Institute séance room – flowers, pieces of jewellery, even a bird. Perhaps taking a leaf out of Harry Price's book, particularly *Regurgitation and the Duncan Mediumship* which had been published earlier in the decade, Fodor and the Institute investigated Mrs Forbes in a similar way to Helen Duncan. But where Helen Duncan consented to being searched, Mrs Forbes always seemed reluctant, in part, Fodor thought, due to the scars left behind after a mastectomy and a kidney abscess. She constantly shifted and fidgeted in her chair in the séance room, arousing the suspicions of the colleagues Fodor invited to watch. Where Helen Duncan refused dramatically to be X-rayed, they somehow persuaded Mrs Forbes to undergo the process, and found a number of small objects concealed about her person that she was clearly intending to produce as materialised apports. By this point, Fodor didn't really believe any more that there was any sort of poltergeist or paranormal entity haunting Mrs Forbes; she was, in effect, haunting herself. He wanted to know how she was producing phenomena and, most importantly from his psychoanalytic approach, *why*.

As this new chapter in the case began, Fodor became more acutely aware of Mrs Forbes' underlying psychological distress. As well as her mastectomy and kidney abscess, he learned of a six-week period of mysterious blindness she suffered in 1929, which went away as suddenly as it came on and, somehow, she managed to keep secret from her husband. Fodor believed this to be a psychosomatic blindness rather than a genuine issue with her eyes, noting that 'she had so much hated the sight of something or somebody that she shut it off by making herself blind.' Indeed, the more time he spent with Mrs Forbes, the more he realised that Mr Forbes was often painted by her as the cause of all her distress. While Fodor could only glean information from what he saw before him, he found no evidence to suggest that Mr Forbes was anything other than a decent man. He was no dashing gentleman, certainly, but Fodor also thought that mild-mannered Mr Forbes couldn't possibly be the root of Mrs Forbes' acute psychological distress.

Fodor was a keen follower and disciple of Sigmund Freud who was, by this time, on his deathbed in London in the final stages of mouth cancer. The paranormal aspects of Mrs Forbes' case were no longer of much interest to Fodor, and he instead threw himself at the challenge of psychoanalysing her. By digging into her past, he identified two key moments in her memory that seemed to continue to haunt her. The first, he believed, occurred when she was a child and was exposed to the smell of human decay when forced to be in the presence of a dead relative. He believed, too, that she had suffered some kind of sexual assault which, while not perpetrated by her husband, was the reason why she often lashed out at him.

The more Fodor turned the case from the paranormal to the psychoanalytical, the more Spiritualists and psychical researchers began to reject him. The Council for the International Institute of Psychical Research, in particular, disregarded his work on the grounds that he had found such a promising and fruitful poltergeist case to be caused by neurosis. With lamentation in his tone he described that they 'made it impossible for me to carry on'. Spiritualist newsletter *Psychic News* similarly disparaged Fodor's handling of the case, and an article from 8 October 1938 posts a rather mugshot-like photograph with the headline: 'Which of These Two Fodors is Right?'[11] They were referring to the way Fodor on the one hand described some of the poltergeist phenomena in the Thornton Heath house as inexplicable, yet insisted that Mrs Forbes was an unwell woman whose own neurosis was leading her to fraudulently produce the strange noises and disturbances. It's quite an insulting headline, especially if their intention was to poke fun at Fodor's theory of the poltergeist as being the result of a 'mental split'; it cheekily suggested that Fodor himself had a mischievous poltergeist that undermines his own work. Their attack on Fodor became so intense, and Fodor was generally deeply troubled by the way his sensitive approach to Mrs Forbes was lambasted by Spiritualists, that he attempted to sue the newspaper for libel.

We've seen many troubling moments in the history of ghost hunting, from Eva C.'s treatment under the predatory gaze of Madame Bisson to the criticism the grieving Lodge family received after the publication of *Raymond*, but this episode stands out because it shows, I think, a particularly nasty side to Spiritualism and psychical research. It's all well and good for them to appropriate scientific terminology and see their practice as highly technical and experimental, but only when one foot remains firmly in the realm of Summerland. Nandor Fodor entered a juicy poltergeist case and, like the *Scooby-Doo* gang, whipped off its disguise to reveal a thoroughly troubled woman underneath. He took responsibility for making her worse by almost encouraging these poltergeist acts, and then abandoned his paranormal investigation in favour of trying to help her. When he saw there was nothing spooky in Thornton Heath, he didn't invent ghosts where there weren't any – nor did he pack up and leave Mr Forbes to pick up the pieces of his distraught wife in order to take on a more promising case. His reward for ethical consideration was to be ostracised from the psychical research community. The only ally who stood by his work was Sigmund Freud, who died shortly after.

In closing his report of Mrs Forbes, Fodor wrote with a note of optimism in spite of the criticism he had faced: 'To the best of my ability, I have analysed the facts and fitted them into a general frame in the sincere hope that this work will help future researchers and will enable them to deal with Poltergeist disturbances in a more intelligent manner than it has been done in the past.'

In this chapter, we've observed how the poltergeist seemed to affect children of a certain age who lived in upheaval or in the substandard accommodation of council houses. Yet this was, for the most part, a British phenomenon, because of the distinctly British situation of mid-century council house schemes. But a parallel trend was occurring in the US. Where British children were being harassed by poltergeists, American families were starting to witness their children talk in tongues, distort their limbs and twist their faces into hideous, uncanny masks. Spirit possession, ushered in by new teenage subcultures that dabbled in the occult, began to tear homes apart.

CHAPTER NINE

Haunted Objects, Haunted People

In the French Alps, close to the Swiss border, is a village called Morzine. Now a popular ski resort, it was once a rural, isolated cattle-farming community of around 2,000 people. At 915m (3,000ft) above sea level, the conditions there are extremely cold and difficult. Before 1925 when Morzine become a tourist destination and with little money to be made in the village itself, the able-bodied men of the parish would leave their families for several months at a time to work in more populous and financially favourable locations. The women and children were left alone to tend to the herds and try to survive each winter and a wind that was said to act 'notably on the nervous system'.[1]

It's no surprise, perhaps, that the citizens of Morzine became possessed.

The events began in 1857, with a 10-year-old girl called Peronne Tavernier. Morzine was a devoutly Catholic community – indeed, there was little else to cling to on the mountainside besides their religion – and Peronne was becoming much more involved in the church through communion and confession. It was after undergoing confession one day in March, however, that Peronne seemed only to move further from God. According to William Howitt's article, written in 1865, upon leaving the church she saw another girl around her age fall into the river, and the incident upset her dreadfully. Later, while at school, she fainted and remained 'as one dead' for several hours afterwards. It happened again a few days later, while Peronne was attending church, and then again and again, occurring with an increasingly alarming frequency. Then, a month later, Marie Plagnat, Peronne's friend, began suffering the same attacks – and was found together with an unconscious Peronne on the hillside. Julienne Plagnat was the next victim, and then Joseph Tavernier, Peronne's brother, followed in June. The mysterious affliction spread through more children, and then through women and any men still residing in Morzine. Eight months after Peronne's first attack, 27 people were exhibiting the same peculiar symptoms.

The victims didn't just faint; they began to display more alarming behaviour which shifted this case away from the realm of medicine to that of spirits and demons. The first children to be affected, Peronne and Marie, started to exhibit the tell-tale signs of possession: erratic gestures, streams of gibberish, convulsions, strange predictions, and blasphemous language and oaths. At one point, Peronne's convulsions were of such intensity that three adult men couldn't hold her down. Julienne's case is more peculiar still: demons seemed to have entered her body through a large and deep cut across her thigh that spontaneously appeared. She said she had been possessed by seven devils – men from Morzine who had recently died.

It was with little Joseph, however, that things seemed to take a particularly sinister turn. Upon succumbing to the illness, he was witnessed running up a pine tree nearly 25m (82ft) high. Then, once he reached the top, he performed a handstand on the top branch while 'singing and gesticulating'. An odd lucidity returned and, as though waking up from a nightmare, Joseph became suddenly terrified at the precarious situation he found himself in. Unsure of what else to do, the panicked bystanders – including his older brother – demanded the spirit to possess Joseph again in order to give him the strength and agility to come back down. The fear vanished from Joseph's face, and he skittered back down the tree headfirst.

After the incident, something changed in Joseph. He stopped eating, and claimed the devils inside him were keeping his mouth shut whenever food was urged upon him. He became weaker and weaker, and died three months after his initial fit. The Taverniers suffered further tragedy as the children's father died, just as Peronne's possessing spirit predicted. More deaths and serious injuries occurred: one woman, possessed by the ghost of a drowned man, declared that she must also die that way and tried, unsuccessfully, to launch herself into the river. As with Peronne's fits, three men had to struggle to hold her back – and only just succeeded in stopping her suicidal campaign.

Something had to be done. It was clear that this was no ordinary disease, and that medicine would not be able to help the people of Morzine. The clergymen of the community were implored to perform the rite of exorcism. They agreed, and gathered the afflicted to begin the procedure. Howitt describes a 'terrible explosion' interrupting the rite, as each member of the community – so keen to have the exorcism performed – resisted with violent convulsions and blasphemies and prevented the clergy from continuing.

News of the demonic affliction plaguing Morzine travelled down the mountain, stoking interest in both the religious-minded and the sceptical. A Lyon physician with an excellent reputation, Dr Arthaud, who specialised in diseases of the mind, was sent to investigate. He didn't last particularly long; his medicines proved useless, and his diagnoses were constantly dismissed under the ever-shifting symptoms and erratic behaviours of the citizens of Morzine. Defeated, he left the parish to continue their babbling blasphemies. His failure only added to the wider intrigue of what was happening in Morzine, and further attention was focused on the small community. Spiritualists and scientists saw Morzine as the epitome of proof both for and against the Spiritualist movement.

Dr Constans, another physician of mental diseases – but of the 'lunacy' variety – arrived in Morzine in 1861, keen to succeed where Arthaud had failed. By this time, there were 120 possessed individuals, not counting those whose possession had already led them to their death. He found that those with which he could hold a conversation all told him the same thing: they were possessed by ghosts, the deceased residents of Morzine, rather than the grand figures of Biblical demonology. Dr Constans was eager to prove this malady a medical one, no different from those of institutionalised patients, but even so he admitted astonishment at the way the possessed moved with inhuman speed, strength and agility, leaping great distances or contorting themselves backwards until their head touched the ground.

But even Dr Constans, under pressure from Emperor Luis Napoleon to cure Morzine of its ghostly disease, seemed at a loss to know what to do with them. Given free rein and whatever resources necessary, he sent in 40 gendarmes and infantry to try to restore order through the threat of military discipline. He warned the parish that those who continued to display erratic and violent behaviour would be punished. In some of the more severe cases, he simply sent them away to various hospitals elsewhere in France to become someone else's problem. As Howitt describes, however, 'failure followed each of his measures'. What was worse, according to Emma Hardinge Britten in 1883, was that the heavy-handedness of Constans seemed only to aggravate the possessed.[2] Such was the frenzy he had unwittingly whipped up that the disease spread to the soldiers assigned to assist him, who began to exhibit convulsions themselves.

Dr Constans left Morzine in a deeper state of chaos than when he arrived, but perhaps the distance away from the demonic growls and unnatural contortions of the parish did him good, because he mulled over his experiences and drew up a report outlining a better approach to treatment. He would not be defeated a second time. The government sent him back again, this time 'armed with the powers of a dictator' and a fresh assortment of soldiers and gendarmes. His patience and scientific curiosity had run out, and he treated the residents of Morzine with severity. Anyone caught convulsing or speaking in strange, blasphemous tongues was immediately deported to an asylum or hospital. Some weren't even given that level of care, and were kicked out of their homes, forbidden to return, with nowhere else to go. At the time of Howitt's report, he declared that it had been four months without any convulsions in Morzine (although he doesn't say whether there was anyone actually left in Morzine to have convulsions).

Howitt's description of the Devils of Morzine concludes with a sympathetic yet damning discussion of the perils of Spiritualism. The residents were not to blame for believing in Spiritualism; it was the chief figures, those who actively sought to spread the word of séances and mediums, who ultimately, if indirectly, brought many lives in Morzine to ruin. Neither in Howitt's report nor in any of the newspaper articles or other writing contemporaneous to the case is there the suggestion that this *was* some sort of virus or physical malady. All conclude with the same idea: a form of mass hysteria provoked by the spread of the Spiritualist movement. Over 150 years later, this theory is still held – Robert E. Bartholomew and Peter Hassall, in their 2015 book *A Colourful History of Popular Delusions*, argue a combination of low levels of education, the rapid rise of Spiritualism in the area, and the community's dominant female population, who passed the time through telling stories (and rumours), laid the foundations for Peronne's initial attack. Moreover, what Howitt perhaps didn't know at the time was that in the early 1850s – recent enough to the outbreak to still be in the public imagination – a young girl in a neighbouring village exhibited a kind of precursor episode to the Morzine possessions, and was taken to Besançon in eastern France to be exorcised.[3]

It wouldn't be right, though, to say that no good at all came from Constans' investigation into the Morzine possessions. While Howitt paints him as a man haunted by failure as those around him were haunted by the dead, a man whose actions became increasingly dictatorial and, in some cases, cruel, he also seemed to recognise the serious social issues which had led to the malady in the first place. As Ruth Harris explains, Dr Constans eventually realised that exiling the afflicted was only a temporary fix.[4] He understood that Morzine was severely isolated, and that the mass emigration of men and the arrival of exciting and popular Spiritualist practices was really at the root of the sickness. He overhauled the social provisions and services within Morzine, using the power given to him by the French government to connect Morzine to its neighbouring towns and villages by reshaping its boundary and developing new roads. He was also remorseful of his initial spate of banishment, and set up proper pensions and stipends for those left behind – although it doesn't appear that he changed his mind to the point of letting back any of the exiled to live permanently in Morzine.

Spirit possession is integral to religion and culture, and in many cases is a revered practice that brings ancestors closer to the living, diminishing the separation between life and death.

In early modern Europe, however, possession became something of a battleground, a phenomenon in which clergymen performed complex investigative duties through tests and, of course, the rite of exorcism.

Exorcisms rose to prominence with the development of Catholicism. Practitioners followed the example of Jesus, who frequently encountered, and exorcised, demons. Perhaps the most renowned case is found in the Book of Mark, where Jesus encountered a man behaving strangely near the town of Gadarenes. The townspeople chained the man inside a tomb, but Jesus saw that the man had wrenched apart his shackles – demonstrating inhuman strength. Sensing the presence of evil within the man, Jesus asked for the name of the demon. 'We are Legion,' the man replied.[5] Jesus turned the demons into a herd of 2,000 pigs, who raced down the hillside in a frenzy and drowned themselves in the sea. Such cases of widespread possession as seen in Morzine were left behind in the nineteenth century. As Spiritualism had a new lease of (after)life following the First World War, trance mediums such as Mrs Leonard and Helen Duncan helped to shift the focus of the dead returning to possess en masse, and instead demonstrated the impact of the individual being plagued by a spirit keen to return to the land of the living in one way or another. Not only that, though, but the rise of modern consumerist culture and individualistic expression, combined with the post-war rise of alternative lifestyles, interest in the occult and the proliferation New Age religions, returned the phenomenon of possession to wholly singular cases. In retaliation to these new ways of living and new kinds of indulgences, the Church framed possession in a different light: a constant fight between the exorcist and the demonic spirit, almost in parallel to the new comic book heroes and their arch-nemeses. While in the previous century the battle for many victims of possession was arduous and complicated, the battle for the individual victim was made all the more dramatic for its one-on-one intensity.

As British ghost hunters struggled to tame the poltergeists that plagued some working-class children, the US was facing hauntings that came from within. In the 1970s and 80s, frenzied accusations of Satanic ritual abuse known as the 'Satanic panic' grew in parallel to the renewed popularity of the Ouija board and sleepover occult rituals. Many believed there to be an epidemic of children in the grip of paranormal forces, demons and malevolent ghosts, who changed their behaviour and appearance beyond what their parents thought acceptable. Cult movies such as 1973's *The Exorcist* only added fuel to the paranoia that children were one tarot card away from crawling the walls and hissing Latin profanities.

Where the UK had had its famous ghost hunters in Harry Price and Guy Lyon Playfair, the US soon saw its own celebrity investigators who emerged to fight these pesky demons: Ed and Lorraine Warren. With Ed as a self-styled 'demonologist' and Lorraine claiming to have clairvoyant abilities, they founded the New England Society for Psychical Research in 1952.

The Warrens are currently being immortalised in one of the most famous modern horror movie franchises involving exorcism and possession, *The Conjuring*. Between 2013 and 2024, the so-called 'universe' of films totalled eight, with titles such as *The Nun* and *Annabelle* featuring as spinoffs from the main storyline. *The Conjuring* films present Ed (played by Patrick Wilson) and Lorraine (Vera Farmiga) as true Hollywood action heroes, pitting their wits and their faith against increasingly violent ghosts and demons who threaten to destroy the lives of innocent, hardworking people. Like private detectives, they step in when the authorities fail tormented families, believing their stories when no one else will. When the evil spirit rears its ugly head, the Warrens discover its history – and its weakness – and, with lots of lightning, shouting in Latin and a big wind machine, banish back from whence it came. The thing about *The Conjuring* franchise is that, of course, the ghosts are always real. Even when *The Conjuring 2*, which dramatises the Enfield poltergeist case covered in Chapter 8, tries to cast doubt over the phenomena as a plot point, it's eventually revealed that the culprit of the disturbance *was* a demon, and not the antics of a teenage girl. There is no doubt for the audience, watching the pallid Nun gurning at the camera, that the events are truly taking place. After an explosive climax, the Warrens defeat the entity and watch, emotionally and physically drained but happy, as the rescued family gather together and begin to heal.

The reality of their investigations is neither so glamorous nor clear-cut as the films make them out to be. Written with Robert David Chase, the Warrens' book *Ghost Hunters*, originally published in 1989, details some of their cases, from murder investigations in which victims spoke to Lorraine from the afterlife, to the banishment of demonic forces from the lives of wayward teenagers. The latter are the most troubling to read. At least with murder cases the damage had already been done; there's a sense that, at worst, Lorraine got slightly in the way as police detectives did their job. But regarding the teenagers, it's difficult to understand the cases from an entirely unbiased viewpoint. It doesn't help that, by way of introduction, Ed declares, 'Many cases of so-called mental illnesses are really the result of possession. Many cases of murder are likewise the result of demonic possession.'[6] Uh oh.

They related the case of 15-year-old Cindy McBain, already painted as rebellious through being described as a 'virgin in the strict sense, though she has "experimented" with a few boys in the same way that she "experimented" a few times with marijuana'. One day, while out shopping, Cindy entered a suspicious antiques shop and bought a Ouija board. At home and in the company of her friend Nancy, Cindy attempted to communicate with the dead, and was rewarded with a persistent knocking sound coming from within her bedroom wall.

Cindy started to lose weight, and this is presented as a spooky detail – a direct consequence of using the Ouija board. At the same time, her parents often heard moaning noises coming from her room. Her mother was a bit slow on the uptake, and it took her a few days to realise that it was the sound of 'people making love'. She barged into Cindy's room, and was astonished to find that her daughter was alone. Before she could come to the logical conclusion, strict Catholic Mrs McBain spotted the Ouija board, and decided that Cindy, 'no longer the daughter I knew', must have invited a demonic spirit into her bedroom.

Cindy became increasingly troubled and emotionally volatile, and her parents sought help from the Church and a psychiatrist. Her father, however, a psychiatrist himself by profession, seemed to push for a serious diagnosis and a few weeks later Cindy was admitted to a psychiatric hospital. From one extreme to the other, Cindy seemed to be torn apart by her parents vying for science or religion to explain their daughter's mood. Mrs McBain, not about to give up on the idea of supernatural forces, consulted Ed and Lorraine Warren. In explaining the particulars of the case, she elaborated that not only had Cindy bought a Ouija board, but had also been 'buying paperbacks on the occult' and together with friends had been casting spells and conducting the rituals described within the books. She also revealed that she had seen a dark shadow drifting into Cindy's room one night, a story she had told no one except Ed and Lorraine. 'Any doubts we'd had about demonic possession', said the Warrens, 'were gone now. This was a classic case of somebody experimenting with the occult and virtually – if not literally – inviting the demon into the household.'

The Warrens visited Cindy in the psychiatric hospital, where they met a priest, Father Elemi, who was already at work on the teenager. The possession gathered speed now that the Warrens were involved; Cindy screamed, swore, foamed at the mouth, and her eyes allegedly changed colour 'from blue to a deep amber colour'. After gaining permission from his rectory, Father Elemi began the exorcism of Cindy McBain. Lorraine described her as jerking 'as if she were being shot with invisible bullets', sometimes screaming in pain, and sometimes lying still as though dead. An hour later, the exorcism was over and Cindy was subdued to the point that her parents believed her cured.

Three weeks later, however, the Warrens explained that the spirit returned to torment Cindy, and this time she was treated by doctors. The Warrens finished their account of this case with a rather harrowing comment: 'Cindy, now a young woman, leads a somewhat normal life – but she is always subject to what her doctor calls "attacks." He calls it mental illness. But we knew better.'

In the mid-twentieth century, mediums seemed to become frequently haunted not by demonic spirits but by ghosts desperate for their help. In the aftermath of the trial of Helen Duncan, materialisation fell out of favour with many mediums. There was too much at stake, and with laboratory equipment 'ectoplasm' was too easily analysed and discredited. Duncan's contemporaries thus began to popularise a different, and in some ways antiquarian, technique of mediumship: the 'direct voice'. This was a form of possession, in which, during a trance state, the medium appeared to be taken over by a spirit personality. It doesn't sound particularly impressive, especially in light of the marvels mediums used to perform, but these spirits would distinguish themselves by having clear voices and behaviour separate to that of their medium, and would amaze their audience through predictions and pieces of ambiguous wisdom.

One of the most prominent mediums of this era was Estelle Roberts (1889–1970). As with many of her contemporaries, she had a simple, working-class upbringing. Born in Kensington, London, she left the local council school at 14 and became a nursemaid. Later, with three children and an ailing husband, she was the sole provider for her family. When her first husband died in 1919, she remarried to Arthur Roberts, with whom she became involved in the Spiritualist movement. Together, they would experiment with sittings and séances in their own home; but soon their work moved into more public spheres.

Estelle Roberts' career as a medium began to develop following a sitting in 1925, in Richmond Spiritualist Church. This was the first occasion in which Roberts' soon-to-be-infamous Native American 'spirit guide', Red Cloud, 'controlled' her.[7] As she relates in her autobiography, she fell into a semi-trance state during a small-group sitting, and was therefore just conscious enough to 'hear' what Red Cloud said. She explains that he spoke through her, and that for a few moments she had no control over her own voice or body. To the gathered party, Red Cloud gave a rather ambitious premonition:

> One day, this medium will be known to all the world. People will come from every country to hear her. Many will be turned away, for there will be no meeting-place big enough to hold all who wish to listen to her. She will never want, nor yet will she ever know riches.

Perhaps this was a form of the New Age trend of 'manifestation'; if you say it with enough conviction, or while channelling the ghost of a Native American, it will come true. As it happened, Estelle Roberts did rise to significant fame, and the king of Greece – at the time enduring exile in the UK – was particularly enamoured by her powers, and often visited to speak for long periods at a time to Red Cloud.

Delivering messages from the spirit of Red Cloud wasn't the only way Roberts worked as a spirit medium. She also demonstrated a technique called 'psychometry', in which an object's past is revealed to a clairvoyant through touch. As though objects themselves record the ghosts of history, they could replay significant events back through the medium's eye. This became increasingly popular in the mid-twentieth century as riskier forms of Spiritualistic performance were abandoned. Not only did psychometry allow the medium to stick to relative safety – it was difficult to shout 'Fraud!' when all the medium did was hold a spoon and say, mystically, that it had been used for eating – but it also reframed the medium's role in society from cabinet curiosity to psychic sleuth. It wasn't so much that people brought mediums their most beloved objects and waited to hear what they already knew; mediums were instead asked to reveal the history of strange and mysterious things.

Estelle Roberts was one who leapt at the chance to use her powers for investigative work. Maurice Barbanell, one of the founders of the College of Psychic Studies and a prominent figure in the mid-century British Spiritualism scene, asked Roberts to come and perform psychometry on an old bed. Tucked away in a Hammersmith antique shop – antiques, Roberts claimed, were something she knew nothing about – the bed looked ornate, yet ornate in an ordinary sort of way, but had a gristly reputation that Roberts claimed to pick up on as soon as she saw it. She hadn't even needed to touch the bed; it was so haunted, so full of bad memories, that they wafted her way almost before she had crossed the threshold of the shop door.

Like a private investigator piecing together clues, she revealed a story that wouldn't have been out of place in the sensational crime novels of the era. She told Barbanell and the woman who owned the shop, who presumably asked Roberts and Barbanell to investigate the bed, that it was made in France and 'occupied the principal room of a medieval ducal chateau'. Watching the scene unfold in her mind's eye, she saw a man writhing in the bed under the effects of poison, followed by another whose murder was so violent that blood from the smashed remains of his head had spattered across the carved wood of the headboard. The third and final murder was a fatal stabbing, and once the culprit had done away with the bed's owner, Roberts saw him frantically searching for 'some papers, which I knew he would not find because they were concealed in a hollow leg of the bed.' Then comes a rather odd statement, in which she implied she had urged the owner to attempt to find the papers, but doesn't hear any more from the antiques dealer. 'But failure to find them', she says, 'did not surprise me at all because, as I have said, it was an ornate bed with much rich carving.' But there are still only four legs on a bed – it couldn't have been that hard to find, unless, of course, the owner had her doubts and didn't want to damage and detract from the price of what sounds like a very nice antique bed.

Not deterred by the lack of proof, Roberts continued to act as a psychic investigator on similar cases, and in 1935 went to uncover the secrets of a haunted house in Surrey. This time, Red Cloud came to offer his assistance. The house was owned by a businessman, who lived there with his secretary and a housekeeper, and since moving in had been troubled by increasingly frequent and disturbing ghostly activity. Spooky faces appeared in the dark, disembodied footsteps followed the occupants around the house, and in the middle of the night the owner's Dalmatian began howling for no discernible reason.

Roberts began with an initial walk-round, accompanied by two male investigators. In an empty gallery that stretched to an unoccupied and abandoned corner of the house, Roberts stopped: she felt a peculiar, strong vibration in the gallery, and was certain that this was where they would discover the house's secret.

There was only one piece of furniture in the long, echoing gallery: a cabinet. Red Cloud emerged, telling Roberts to focus her attention on the cabinet. No sooner had she heard his advice than a monk appeared beside the cabinet, desperately searching for something. The following exchange occurred:

> 'Can any of you see this figure?' I asked my companions. None of them could.
>
> 'Have those of you who live here ever seen a man wearing a cloak and cowl?'
>
> 'Yes,' the secretary and the housekeeper answered almost in concert.

Roberts described what she could see; even now, the monk was still looking for something in and around the cabinet. The owner was puzzled, and told Roberts that there was nothing inside the cabinet – of that, he was sure. The monk, seeming to hear the owner, suddenly ceased his search and disappeared.

Using a combination of Red Cloud's direct voice and psychometry, Roberts revealed the story behind the haunting. The monk that she, the secretary and housekeeper had seen was the brother of a former owner of the house. Once again, it's a gruesome story: the monk's brother had been strangled, but before his murder had – like the owner of the haunted bed – hidden some papers in the cabinet. When the monk died, he returned to the physical plane to continue searching for the papers, but always in vain.

Roberts was asked by the owner what made the papers so special that a man was murdered for them. The answer is, oddly, not much: she only knew the contents were written in Hebrew and Greek, 'sealed with a great seal'.

Alongside possessed people, then, there are possessed *objects* – possessions that a ghost can inhabit, causing it to move around or curse its owner. Such belief has been recently popularised by, again, *The Conjuring* franchise. The Warrens kept a museum of allegedly haunted objects in the basement of their house, which features in some of the films as a kind of epilogue, where Ed or Lorraine add each film's most prominent spooky prop to the shelves. Among these items is Annabelle, a large rag doll which tormented a student nurse in the 1970s. Dolls, by their very nature of being a bit creepy, seem to have become a cultural magnet for the paranormal in recent years.

I've mentioned before about how paranormal investigators constantly evolve their vocabulary and ghost-lore, and haunted objects have become a considerable part of this. Alongside their abundance of equipment, ghost hunters may bring what they consider to be a haunted object with them to help build up spiritual energy, and even sometimes stumble upon spooky dolls left behind in locations. Recently, too, some ghost hunters warily discuss a spirit's ability to 'attach' itself to them and follow them home, essentially turning themselves into a haunted object. They will protect themselves with rituals, and verbally warn a ghost not to stay put.

Harry Price brought back a few souvenirs from Borley Rectory. The investigations were surprisingly abundant in the appearance of tokens and objects. A coat that belonged to no one was suddenly found hanging on the back of a door without explanation, and a nail file and a gold wedding ring materialised in rooms in the rectory – again without being claimed by anyone. The latter two are still held in Senate House Library, and while I was there I requested to see them. The nail file is small and slightly rusty, and the wedding ring was a plain, hallmarked band. Both seemed unremarkable, but in the context of Price's investigation they feel like they carry so much history that, in a way, they are haunted. Also among the relics from Borley was the fragment of skull that had been dug up during Price's excavation of the cellar and grounds.

In September 1943, Price took his collection of spooky objects to an antiquarian photography studio in London renowned for being employed by the Royal Family and for their careful treatment of rare artworks. There, he handed the objects over to the company's director, Mr Cooper, who seemed very bemused by what Price was asking him to photograph – but nevertheless accepted what was probably a very handsome fee. Price remarked in *The End of Borley Rectory* how Mr Cooper 'knew nothing whatever about psychical research, and – though he was not rude enough to tell me so – probably cared less.' As I've just mentioned, Messrs A. C. Cooper, Ltd. was a well-established and highly regarded company employed to photograph the finest and rarest works of art that passed through London. Their photographers didn't simply drop things. But immediately as Harry Price handed the skull fragment to Mr Cooper, he dropped it. It broke into four pieces, and left Mr Cooper much distressed. In all his 25 years photographing priceless antiques, he had never damaged an item. He reassembled the pieces with specialist glue and photographed the skull fragment, and Price left Cooper with the rest of his collection while air raids continued to threaten them from above. The next day, Price returned to pick up his Borley relics, and Cooper couldn't wait to be rid of them. Disaster after disaster had occurred in his studio; a priceless oil painting crashed to the floor from its easel, followed by another painting on a different easel, and then Cooper ruined two batches of photographs by neglecting to prepare the camera properly. The final straw involved an old clock hanging on the wall that hadn't worked for 10 years. Suddenly, it began to tick. All in one afternoon. Cooper pushed the relics and the completed photographs into Price's hands and bid him never darken his doorstep with haunted objects again.

This case study is absolutely fascinating, and shows our reaction to haunted objects. Perhaps because they're tangible things we can hold, we respond more acutely to them in relation to our own superstition than when we're inside an alleged haunted house. Our relationship with objects is, I think, more intense than with environments. Or, at least, it's different – more focused. In a haunted house, we don't know where to look first; the ghostliness of the place is all around us, formless. A haunted object attracts our gaze and holds it in an almost primal standoff between a threat and the threatened. On the other hand, I think it's crucial that Price noted the current spate of air raids over London during the time he visited Mr Cooper. He wrote about it rather nonchalantly, but it was clearly a dangerous time to be living in London. My suspicion is that Mr Cooper was rattled by the dread that his building, his livelihood, his life, might be blown to pieces within days. That was why he was clumsy with the skull and the films he ruined. Perhaps, even, the blasts that had shaken London had caused buildings and floors to become a little unsettled; the paintings falling off their respective easels, the clock that began to work after 10 years, may simply have been shaken up, too. Or Harry Price really did bring ghostly energy into Mr Cooper's studio. Nevertheless, despite his being a sceptic with no interest whatsoever in the paranormal, and despite being someone who had likely photographed antiques and artefacts with an incredibly dark history of being plundered by the British Empire, the objects brought from a haunted house rattled Mr Cooper more than any enemy bomb.

The most notorious haunted object is the Ouija board. As mentioned earlier, these saw a boom in popularity during the Satanic panic of the 1980s, and led poor Cindy McBain to encounter the Warrens. As Spiritualism took the world by force, its commercial potential was noticed by those outside of the séance room. Companies found ways to bring the spirits into everyone's home. You no longer needed to spend an extortionate amount of money to have a private sitting with renowned mediums such as Margaret Fox or Eusapia Palladino. Anyone could be their own medium, and all they needed was a simple, mass-produced tool: a Ouija board.

The Ouija board was invented by a group of Spiritualists in Ohio towards the end of the nineteenth century as a speedier way to have spirits communicate than rapping out the alphabet constantly. Their design – an alphabet inscribed onto a board with a teardrop-shaped wooden planchette that the spirit could move to a certain letter – was taken up and patented by Elijah Bond in 1890. It was later mass-produced by Hasbro as a 'board game', and is still available to buy under that marketing today. My grandma apparently liked to have a go with her two sisters, and once roped my mum into joining them; mum told me the glass they used in place of a planchette did move, but she insists that her mischievous aunt was behind it.

Ouija boards seem to have gained a reputation for being cursed and haunted objects straight out of the box. Urban legends surround them, warning the inexperienced to never dabble with one despite them being manufactured by the same company that makes the board game Monopoly (which I would argue is much more cursed and always left me and my sisters requiring an exorcism). Of course, Ouija boards now have their own set of Amazon reviews. One is particularly interesting: the reviewer had wanted to try a Ouija board, believing their house had gone through a period of being haunted a few years prior. They said that it worked, but don't go into specific details. What they do give, however, is a warning. After using the board, the strange phenomena began again, as though the spirit had been summoned back. The reviewer warns future buyers to be careful, and to respect the restless spirits of the dead. They gave it four stars.

I'm fascinated by the way a tangible object can seem to provoke feelings of dread in people. Ouija boards are the most prolific, but dolls also have a reputation – recently helped in part by Annabelle from *The Conjuring* franchise – of being prone to ghostly possession. We've seen ghosts take many forms so far: ectoplasm, materialised spirits, translucent spectres, electromagnetic waves. Some of these appearances are more plausible than others, but somehow the thought of a ghost inhabiting a little doll forever is a stretch too far for me. How do they choose which doll? Why does the doll have to be creepy? If I were to haunt a doll, I'd at least pick something cool like a Japanese Gundam robot.

Confronting my own scepticism, I decided to acquire a haunted doll of my very own. It turns out that there is a roaring trade in such accursed items on eBay, and really I was absolutely spoilt for choice. There were rag dolls like Annabelle, modern dolls, 90s TY Beanie Babies (that's one way to get rid of them, I guess), and traditional dolls in frilly hats and dresses. There were some quite peculiar objects, too, including photographs, candles, necklaces and even a pair of haunted leggings. When I die, I'd hope not to possess something so close to a person's sweaty gym thighs, but that's just my personal preference I suppose. It seems to be quite a lucrative operation, too, with some of these haunted dolls fetching between £50 and £100. Generally, the more creepy and Annabelle-like they look, the higher a price they demand.

I spent a while looking for the right doll; not too big, not too creepy. I even got a group of students to help during a mid-lesson break (they wanted me to get a possessed 30cm (12in)-tall Aragorn doll from *The Lord of the Rings* series, which the eBay seller promised was haunted by a somewhat lecherous male ghost who liked to whisper sweet nothings in your ear at night). In the end, I settled on a small mid-century doll whose advert described her as being haunted by a ghost called Bev, a woman who in life enjoyed cooking and looking after injured wildlife. I thought that sounded cute.

My problem is that I become easily and wholeheartedly attached to certain objects, almost as though I haunt them already. I'm not a hoarder by any means, but there are a certain number of things I own which I would genuinely grieve over if anything happened to them. Some of them have real sentimental value, of course, such as the little wooden house my grandpa brought back from Innsbruck when he was in the Merchant Navy during the war. Others could easily be replaced, like the mug I have my morning tea in or the big yellow blanket I have at the bottom of my bed – but a replacement wouldn't be the same. It's at its worst when someone gives me an object; I feel compelled to keep it then, because I think that getting rid of it is the same as stabbing the giver in the back. Plus I think it's nice to give things to people. You could give me a bent cap off a beer bottle and I'd probably keep it forever as something to remember you by. I bought Bev and became immediately emotionally invested in her harrowing journey through the British postal system.

When Bev turned up, wrapped in bubble wrap and Sellotaped into an old Morrisons bag, I carefully propped her up on my desk. As I moved my hand, she blinked her lids closed with a soft clack, and I gasped. I wasn't expecting her to blink. I realised, then, that there was a part of me that was nervous to own such a thing. The folklore around haunted dolls, to my surprise, did rattle my scepticism.

Bev is around 25cm (10in) tall, dressed in a green skirt and red gingham smock with matching hood. Shiny blonde waves peek out from under the hood, and her blinking blue eyes are always fixed on something in the corner. I think she has a look of Doris Day about her; she seems to be from the 1950s, so perhaps the actress did influence Bev's design. Also included in the package from the eBay seller was a laminated information sheet about Bev's history and personality, written in the same way animal rescue centres describe their residents as 'easily overstimulated' and 'not to be housed with young children'. It describes her as 'a kind and generous lady that died when she was 77 years old'. During a séance with the seller, she told them that she had lived in Nottinghamshire – I'm sure she's now thrilled to live with me by the sea – with her husband Douglas. She was an active church-goer and often raised money for charity including, says the info sheet, 'orphaned foxes'. Weirdly specific, Bev, but fine. The sheet goes on to recommend keeping Bev in the kitchen because she 'loved baking cakes', and the seller had been seeing strange phenomena there whenever they baked. Why Bev decided to haunt a doll that looks like Doris Day is 'unknown'.

I took Bev into my office at work, rather than keep her in my kitchen. There were two reasons for this: firstly, my mum didn't want to visit my flat for the weekend with a haunted doll. I was also starting to chalk up odd coincidences to Bev's inner phantom. At the time, I'd only recently moved into my new flat and still wasn't used to the various noises of the building, so I was sleeping with ear plugs. I'd been doing this for a few months and had never lost a pair, but as soon as Bev turned up I lost two in the space of a week. Not only that, but they disappeared completely. There are only so many places ear plugs could be in my flat – the space isn't all that big – and I searched everywhere. Since the only other explanation is that I ate them in my sleep, I think I'd rather pin it on Bev ectoplasmically dematerialising them. At the same time, something else was happening. There are four integrated ceiling lights in the spare room where I write; the one right above my desk, above where I'd perched Bev, started to fizzle, and then a second light started to blink a few days later. When I took Bev to my office, the light bulbs worked fine again. My superstition was starting to get the better of me.

As far as I'm aware, nothing odd has happened at work now that Bev is a Visiting Fellow. She is currently leaning against the wall, her feet touching a scrap of white paper in which I've placed two pennies as trigger objects. The pennies haven't moved outside their traced outlines but, of course, that's not to say Bev doesn't move them in the night and move them back again. Perhaps if I put a crisp £5 note there instead, she'd be more inclined to nick it and head off for a cheeky drink at the Student's Union bar.

It's hard to write about Bev and not imbue her with personality. This is the effect I've found she has. I've only to say to people that she's haunted, and they instantly react with either delighted fascination or horror. They assign life, or rather afterlife, to her more readily than they would if I didn't say she was haunted. When I told my boss that I'd brought a haunted doll to my office, she sighed with long-suffering acceptance of my weirdness and told me not to open a hell mouth in the department. I took Bev into my Writing Horror class one day, and extended her little pink hand for my students to shake between thumb and forefinger. Some students *very* firmly declined; others gingerly touched her, still leaning away slightly as if expecting her to start cackling or revolving her head. A couple of colleagues will give her a wary, side-long glance as they step into my office to ask me something. Bev is a perfect litmus test for gauging people's superstitions, I've found. She seems to weaken people's scepticism, even my own.

But still, I was sceptical. The more I read about paranormal phenomena and ghost hunting, the more confused and conflicted I became. There are so many anecdotes, so many investigations, and none of them seem to come to any agreement about what ghosts are or how they manifest – so how can I be expected to believe that the soul of a human called Bev from Nottinghamshire inhabits a 25cm doll bought on eBay? Perhaps, though, I simply wasn't investigating her properly.

Since my head of department would disapprove of me conducting a séance on campus, I briefly brought Bev home again, tucked in my backpack. I'm pretty good at baking, so I balanced her against my microwave while I made millionaire's shortbread. I felt Bev's bright blue eyes on me all the time, and imagined her yelling at me not to stop stirring the caramel or it'll burn. The end result was delicious and decidedly not haunted. Ghost hunters have done so many strange things in their pursuit of ghosts, but I'm not sure how many have baked treats in order to communicate with the dead. As a séance, though, it left me no closer to experiencing paranormal phenomena. It did give me a cracking sugar rush, though.

I decided to do things the traditional way. A lot of modern paranormal investigations see 'protection' as a necessity, and the Satanic panic of the 1980s had firmly established superstitious warnings against using a Ouija board. But I was getting to a point in my research where I felt as though I was becoming more and more disappointed, more and more sceptical. So that evening I did a séance on my small, round dining table. I kept the green polka-dot tablecloth on because I thought Bev would probably appreciate my mid-century décor. I propped her up against a selection of the Spiritualist and occult books I'd been collecting over the last few months, including Dion Fortune's *Psychic Self-Defence* and Sir Oliver Lodge's *The Survival of Man*. I put Mary Berry's *Baking Bible* at the bottom of the pile, too, which is definitely not what her book was intended for. I turned on my lamp, closed the curtains, and I sat. I had a go at automatic writing, and tried to let my mind go blank as I kept a pen poised above a notepad. I asked Bev to give me a message, and the self-consciousness of talking to a doll alone in my flat only made my scepticism rear up in defence. I sat, and sat. I heard my neighbour upstairs get home from work and start pacing around his kitchen. The train rumbled past. I sat. I sighed, and doodled a square on the paper, but I don't think this was any kind of message from Bev unless she was trying to communicate that I should do a traybake next.

Giving up, I turned the pad on its side and made a crude Ouija board. I fished around in my cupboards for something small I could use as a planchette, settling on an egg cup. I upturned it and placed it in the middle of the notepad, loosely resting my finger on the top. Again, I tried to relax and clear my mind. I asked if Bev was there. Nothing. I asked if anyone was there at all.

I thought about my grandma. I looked at the large, framed photograph above my sofa of Fred Astaire mid-dance in *Roberta*, wondering if she'd ever seen it. The question formed in my head, and I found that I half-expected the egg cup to move to 'yes' or 'no'. Still, nothing. Then I made myself sad, heightened by an embarrassed sense that I was being ridiculous and weird, and I cleared the table and the next day I took Bev back to my office on campus.

As we've seen in this chapter, paranormal investigations changed in the late twentieth century in relation to the Satanic panic and the United States' preoccupation with demonic forces. Possession has been a part of religion for centuries, and in some respects was an accepted mode of communication in Spiritualism through materialisation and the direct voice. The way that ghosts can haunt objects and people, not just their environments, is an aspect of modern ghost lore that has turned the practice of ghost hunting from observation to intervention.

CHAPTER TEN

Who You Gonna Call?

When I was 13 years old, I turned on the television one night to be faced with an eerie green scene. A woman with hair and face the colour of glowing lichen peered upwards, uncomfortably close to the camera, while the background faded into black behind her. Something startled her, and she screamed. Loudly. Then the moment of her shock was replayed, as was her scream, several times.

This was my first experience of *Most Haunted*, a ghost-hunting reality television show that began airing in 2002 and quickly became something of a British institution with an international legacy. Its first episode took place at the imposing fifteenth-century Athelhampton Hall, a privately owned manor house whose married owners, Patrick and Andrea

Cooke, were unsettled by paranormal activity from a number of alleged spirits including, it was believed, the spirit of a monkey forever in search of its long-dead master. Presented by Yvette Fielding, whom many viewers will remember from another British institution, the BBC children's programme *Blue Peter*, she introduced the concept of the show with gentle earnestness. Walking through the walled garden in a black leather trench coat that showed how much *The Matrix* was a current public obsession, she called *Most Haunted* 'one of television's most extensive investigations into the paranormal' which guaranteed 'no tricks, no camera illusions and no practical jokes.'[1] The opening minutes of the programme implored the viewer to take this broadcast investigation seriously, while also taking steps to predict and deflect immediate concerns from sceptics. Fielding emphatically tells us, 'If we find nothing, you'll see nothing.'

I was struck by this opening when I saw it for the first time recently. I started watching *Most Haunted* at the height of its popularity in around 2006, and had therefore missed the early episodes. As we'll see later in this chapter, this sincere promise of no trickery or fraudulent phenomena was quite short-lived, which makes the first episode all the more intriguing. After giving an introduction to the history of the house, Yvette candidly shows us the rest of the crew, who smile and laugh and wave at the camera as they unpack camera and sound equipment. We're also introduced to medium Derek Acorah, the soft-spoken Liverpudlian, who would for a long time be the most notorious part of *Most Haunted*'s structure. Paralleling the early days of the Ghost Club, the crew and the young owners of Athelhampton Hall sit around the fire asking questions about their idea of the paranormal. Suddenly, the Cookes' dog becomes transfixed by something in the hallway beyond the hearth, and Derek claims to see the ghost of a monkey which the dog is defensively staring down (although, we're told, Derek has been given no prior information about the house's alleged hauntings). The entire first episode is endearingly simple and amateurish. Their parapsychologist Jason Karl infrequently wafts an EMF meter around a few doorways, but otherwise the equipment is simple. It involves watching and listening in much the same way as we were taught in the ASSAP training day – no Ouija boards, no flashing equipment. The owners remain with the crew, and there's a sense that it's more about corroborating their experiences, as the ASSAP seeks to do, than stirring up trouble for the sake of entertainment. Indeed, in the end credits of the episode there's a small clip of the owner begging Derek not to disturb the ghosts and make things worse – which he humbly promises not to do.

The 'evidence' in this first episode is slight, and is all the more effective for it. While all accounted for and talking in a bedroom, the crew are suddenly disturbed by a repetitive creak and thud. Yvette's face quite visibly pales as she realises it's the heavy wooden cot, as old as the house itself, in another room. The camera jerks as they rush to the source of the sound, and owner Andrea Cooke, one of the first on the scene, claims that she saw it deliberately and slowly rocking before coming to an abrupt stop as they clamoured into the room. She is clearly upset and shaken, repeating under her breath that she has never liked that cot, and is seen blowing her nose behind Yvette as though she has been frightened to tears by what she has seen. The floorboards are noisy, and as Jason Karl explains, no one could have rocked the cot and moved away without the sound of their footsteps being caught on camera. Later in the evening, having moved the creepy cot into another room, Yvette and Jason watch as eerie flashes of light, like electrical sparks, spontaneously appear in quick, flickering patterns in the air around the cot.

This is the sum total of evidence gathered in the first episode of *Most Haunted*. The movement of the cot, while freaky, could have been caused by settling wood, and the sparks of light could be a fault in the camera. Or, they could be paranormal phenomena. *Most Haunted* made no conclusions in this first episode. What's particularly interesting, too, is how simple and amateurish the investigation is. Yvette becomes increasingly nervous, and spends the second half fidgeting, with her fingertips against her mouth. There's a sense that the crew weren't, in fact, expecting anything to happen at all; the fact that they did catch a couple of strange incidents leaves them subdued as the night goes on. I was pleasantly surprised by how unassuming this first episode was, considering what the show would soon become.

The format of *Most Haunted* was similar to another British television programme that had nearly traumatised a generation of people on Halloween night 10 years before: *Ghostwatch* (from which the 1999 found-footage horror film *The Blair Witch Project*, another of *Most Haunted*'s influences, took some inspiration).[2] Hosted by well-known presenter and chat-show host with a reputation for serious, in-depth conversation, Michael Parkinson, the programme was delivered to viewers as a live broadcast in which a television studio 'hub' linked up with a group of investigators at a council house in London. The house was occupied by a single mother and her two adolescent daughters, who, like Janet in the Enfield poltergeist case, were the epicentre of a violent ghost whom the girls called 'Mr Pipes'.

The BBC was no stranger to paranormal programming. While this was the first television offering of its kind, the broadcasting company had an extensive history of offering ghost-hunting programmes over the radio. Members of the BBC had undertaken investigations at Borley, and Price was continually sending the corporation lectures and talks for potential broadcast. In 1934, the BBC ran a series of talks from Spiritualists and psychical researchers under the title *Inquiry into the Unknown*, including MP Sir Ernest Bennett and Sir Oliver Lodge giving talks about haunted houses, psychical research and spiritual survival. It was so successful that the speakers received around 3,000 letters from interested listeners.[3] The BBC even held a prototype of *Ghostwatch* and

Most Haunted two years later, on 10 March 1936. Called *The Broadcast of a Haunted House*, it was a live programme over the radio in which a crew (including Harry Price, needless to say) conducted a ghost hunt and reported their findings in real time to an enraptured audience. The broadcast began by outlining its purpose, pre-empting Yvette Fielding's words in the very first episode of *Most Haunted*: 'We're going to try to give you a chance of hearing how a scientist sets about an investigation of this sort.'[4] Presenter Mr Grisewood outlined the manor house's history before describing the floor plan to the listener and, exactly as *Most Haunted* does, introduces the rag-tag gang of presenters, sound engineers and other crew members on site. There were brief vox-pops of eyewitness accounts, and then Grisewood ended his introduction in a manner which *Most Haunted* would uncannily parallel 70 years later: 'If nothing happens, and I don't suppose anything will – we've not proved anything; we've not proved that there are ghosts or that there aren't.' This innovative form of radio broadcast proved that a live ghost hunt, broadcast to the nation to vicariously encounter ghosts alongside presenters, paved the way for *Ghostwatch* in 1992.

Ghostwatch began as light-hearted fun. The go-between for the studio and the house was actor and comedian Craig Charles, whose unbridled enthusiasm many recognised from the British sci-fi sitcom *Red Dwarf*. It all seemed like a bit of an adventure. The team interviewed Pamela Early and her daughters, Suzanne and Kim, who described in sullen tones the sorts of activity that had been bothering them – the usual things such as knocking sounds, being pinched and having bedclothes torn away while they slept. In the studio hub, Parkinson took calls from viewers and interviewed parapsychologist experts to see what they made of the story.

As the programme progressed, little by little, strange things started to take place. Craig Charles wasn't quite so bubbly any more, and his co-presenter Sarah Greene (who, like Yvette Fielding, is a former *Blue Peter* presenter) began to show the first signs of trepidation. The director occasionally cut to the studio, where Parkinson and his guests sat in awkward silence before trying to resume proceedings.

At one point, as Greene and the camera crew were in the girls' bedroom, the camera panned quickly to show a tall figure standing near the curtains. The phone in the studio went wild with calls. Parkinson began to lose his signature composure. Things became increasingly violent in the house, and seemed to be focused on the door under the stairs that led to the basement. It was revealed that Mr Pipes was a man named Raymond Tunstall who hanged himself in the basement after being convicted of paedophilia. He spoke in nursery rhymes through a haunted-looking Suzanne, who crouched down by an armchair as picture frames leapt from the walls, just as Janet's voice had become gravelly and strange as the poltergeist possessed her in Enfield.

The studio went haywire, with cameras short-circuiting and lights flickering. In the house, all was chaos and screaming, and the door to the basement spontaneously opened, dragging poor old Sarah Greene inside and shutting her in. Amid the smoke and confusion back in the studio, Michael Parkinson wandered alone in a daze towards the camera, mumbling about being deserted by the crew and audience, bewildered by what had just transpired. He came so close to the only working camera that you can see only his chin and shoulders. It was then that he began to speak in nursery rhymes, in that same deep, whispered voice that spoke through Suzanne a minute earlier. The camera dramatically cut out.

The British public were left horrified, believing their beloved Michael Parkinson had been possessed by the ghost of a paedophile and that Sarah Greene, whom many viewers would remember from their childhoods spent watching *Blue Peter*, was trapped in a basement to meet some sort of equally terrible fate. Complaints and frightened calls came in thick and fast to the BBC – allegedly including from Parkinson's poor mother – and the parents of a young boy who killed himself after hearing the central heating pipes banging.

The programme was, however, a work of fiction.

I watched *Ghostwatch* for the first time while researching this book, already having read up about it and its controversy, and I could fully understand how it misled the public so disastrously. A colleague who'd watched its original broadcast told me how terrifying it was to witness that night. The phone calls Parkinson received in the studio were from actors, which were delivered in a rather wooden and melodramatic fashion, and that's about all that would really make a viewer question the reality of the programme. Parkinson, Charles and Greene had all been involved in serious journalism before, and their part in the show seems entirely trustworthy. The special effects, too, such as the objects that whizz violently around the room, are combined with shaky camera-work and chaos to make the incidents feel genuine. It really draws you in, and the dramatic end in which Parkinson hisses nursery rhymes in an empty studio is sudden and shocking. It was unlike anything that had ever been shown on British television before, but its popularity and publicity showed that the viewing public did have an enormous interest in ghost-hunting programmes. It laid the foundations for *Most Haunted* and its international copycats.

By the time I watched my first episode around 2006, *Most Haunted* had evolved considerably from its simple, humble beginnings. As we saw earlier, it had begun as a documented vigil, not worlds apart from what we did at the ASSAP investigator training day; it involved keeping watch in rooms renowned for activity, and recording, analysing and discussing any phenomena experienced in a practice that followed the guidelines of established ghost-hunting societies.

But as its early episodes proved to be successful, the format and content of *Most Haunted* became increasingly elaborate, straying far from its simple beginnings. We've seen time and again how, when a medium, poltergeist case or spooky house starts to garner attention, a kind of performance anxiety settles over those closest to the haunting. Florence Cook's séances grew too elaborate for her to control, Mrs Forbes became increasingly desperate to impress Nandor Fodor with her apported objects, and Janet Hodgson attracted the suspicion of investigators as the poltergeist was asked to demonstrate miraculous phenomena. The same happened to *Most Haunted*; it appears to be a universal effect that dooms anyone who claims to produce evidence of ghosts. People want more – they want better.

This is, I think, also the fault of framing *Most Haunted* as one would a horror movie. *The Conjuring* franchise didn't grow in popularity because each film is tamer than the last; with horror sequels you have to go bigger, scarier. *Most Haunted* couldn't keep producing the same subtle, genuine evidence and maintain the audience's interest. Sooner or later, viewers would want more or they would stop watching. The crew in the opening episodes did their jobs, but chipped in as witnesses if anything happened to them. Later, however, they became protagonists. The team expanded with an ever more bizarre mix of people including, at one point, Yvette's makeup artist, all experiencing their share of phenomena (accompanied by eerie background music that is, fun fact, the same thematic score as the classic 2001 action film *Lara Croft: Tomb Raider*). Just as any good horror story begins in relative brightness and ease to set up a greater contrast with the characters' impending doom, so too did *Most Haunted* operate by making its opening segment light-hearted, demonstrating the team's enthusiasm and eagerness to begin the investigation that would soon see them screeching in fear.

As daylight waned, the team would switch to their night-vision cameras, which created the sickly green glow I saw when I was first drawn into the world of *Most Haunted*. The evidence they recorded soon became exceptional, and unlike anything that had been documented before. There was never a dull episode; time after time they caught shadows and orb-shaped 'light anomalies' whizzing across the camera, balls were thrown back to them down empty corridors, doors slammed shut, whistles echoed from disembodied lips, and crew members were scratched through their clothing.

Rather than keep up with modern parapsychology, *Most Haunted* seemed to regress through the history of ghost hunting. This is fascinating to me, especially since its early episodes seemed to present the programme as one that would maintain a rational, forward-thinking approach. But there's a reason why sensational clairvoyance and séances drew a crowd. They're entertaining – more so than watching people stand silently in a room, waiting for something to happen. As *Most Haunted*'s popularity grew, it started to embrace techniques from Spiritualism's heyday; they often used a Ouija board, did some automatic writing with a planchette, screamed as a table tilted violently under their fingertips, and, in what became a staple phenomenon of the show, had their questions answered through knocks in floorboards and ceilings just as the Fox sisters claimed happened to them over 150 years before. They seemed to be able to connect to ghosts without any effort at all – even during the daytime walk-around, ghosts would pop in willingly to have a lengthy chat with Yvette about who they were and how many spirits still roamed the building. A particular phenomenon, tapping which mimicked the beat of a human heart, started to occur frequently. It happened so often, in fact, that Yvette Fielding quite often talked to eyewitnesses and other members of the crew over the top of the sound – it was no longer remarkable.

Their evidence became increasingly bombastic, with cast members becoming possessed, vomiting, or pushed by invisible hands. There was also the matter of their original spirit medium, Derek Acorah. In the first episode, in what he claimed was a first for him, he seemed to become influenced by the monkey that was said to haunt the manor – miming the mannerisms of the animal. His signature move, though, was to channel the spirits in a croaky Beetlejuice-esque voice as many mid-century mediums such as Estelle Roberts had done. Then, in 2003, during one of *Most Haunted*'s special live broadcasts, Acorah went into a trance and began to spout spooky double-entendres such as 'Mary loves Dick' and 'she protects him; she covers him up'. There's a pause as the crew wait for the focus to return to the central studio, but, unbeknown to them, their camera is still live – their serious expressions falter and they burst into laughter. While this was brushed off as simply reacting to Acorah's genuine trance, there's something about the moment that makes it seem as though the crew are laughing at a shared joke.

This incident was only a precursor to the bombshell allegations *Most Haunted* was about to face. In 2005, in an episode investigating Bodmin Gaol, the team's new parapsychologist – who always tried, rather in vain, to keep everyone level-headed and rational – Dr Ciarán O'Keeffe, set a trap to expose Acorah which was something straight out of the Mollie Goldney playbook. He planted 'information' about a prisoner named Kreed Kafer, a South African man who never existed. Duly, Acorah found the information, and manifested Kafer's spirit during filming. Not only was Kreed Kafer a product of fiction, but his name was an anagram of 'Derek Faker'. Moreover, in 2005, the UK's regulatory body for broadcasting, Ofcom (the Office for Communications), established a new Broadcasting Code which was used to assess all programmes shown from the July of that year onwards. The new code outlined that programme producers' specific duty was to 'promote media literacy', giving audiences 'the tools to gain a greater understanding of the context in which content is broadcast'.[5] By December 2005, it had already assessed *Most Haunted*.

According to the report, 11 people had complained about the programme, thus prompting Ofcom to scrutinise the format. In summary, the complaints were along the lines that the programme promoted 'fraudulent practice', that viewers were being 'deceived' into thinking evidence was genuine, and that 'there could be potential harm to susceptible or vulnerable viewers as a result'.[6] Even as the programme was being examined by Ofcom, more complaints were coming in. The saga was reminiscent of so many of the other cases we've seen throughout this book – from the trial of William Mumler to Helen Duncan's exposure – showing how, centuries apart, there will always be people for whom their scepticism is transformed into impassioned concern by the success and popularity of those who claim to have been in touch with the other side.

Ofcom's inquiry into *Most Haunted* focused on what constitutes a paranormal investigation. Responding to Ofcom, production company LIVINGtv admitted that their 'experiments' couldn't be replicated in a laboratory setting, which, indeed, undermined the legitimacy of what they were painting as a scientific investigation through the presence of O'Keeffe and his various pieces of equipment. Nevertheless, they defended the programme as 'indisputably retain[ing] an investigative element'. Ofcom, in turn, confessed that its role as a regulatory body focused solely on whether programmes uphold its codes, not whether paranormal evidence is legitimate or fake. There's a sense in the report that Ofcom, faced with this new kind of reality television programme, simply didn't know what to make of it. Ghost hunting is such a complicated business, rife with moral ambiguity and questions of scientific authenticity, and Ofcom came face-to-face with the same challenge we've been trying to unpick throughout this book.

However, the regulations at the time stated that any demonstrations of mediumship and other paranormal practices broadcast must be accompanied with a clear disclaimer that it is purely entertainment. Before each subsequent episode of *Most Haunted*, a message now appeared to cover LIVINGtv's back against a less lenient future Ofcom investigation: 'This programme is for entertainment purposes only.' New episodes of *Most Haunted* ceased to be shown on traditional television from 2019, and, instead, the crew began to upload episodes to an official YouTube channel which, as of June 2024, had over 125,000 subscribers. As soon as an episode starts, a new message is displayed: 'The following programme is definitely NOT for entertainment purposes only. We stand by the legitimacy of this investigation.' It is a clear tongue-in-cheek retort to the limitations imposed on them by Ofcom, which don't apply in the Wild West that is online video streaming. Yet, for all that the *Most Haunted* team now stand by the legitimacy of their investigations, the evidence is as bonkers as ever. In a 2023 video of Bishton Hall, Staffordshire, chairs are violently thrown around a room to much yelling and screaming – but, crucially, the camera is always turned away, only jerkily coming to focus on the chair after it has landed.

Most Haunted's format was so simple, so addictively formulaic, so lucrative, that it didn't take long for other production companies to create their own version. Ofcom report be damned; there was something so undeniably entertaining about the show that kept people hooked. As well as *Most Haunted*, I used to watch Welsh-language channel S4C's *Ofn* (meaning 'fear'), which was, bizarrely, part of its afternoon run of children's programmes. It followed the same format, but stuck to Welsh locations and its two presenters would work with a group of tweenagers long past their bedtime to investigate any spooky goings-on. Its reach has been wholly international, with copies of its format appearing in Mexico, the Ukraine and the US. In 2004, American channel Syfy began its own series of *Most-Haunted*-style offerings, beginning with the popular *Ghost Hunters* (2004–16). But it was the Travel Channel's *Ghost Adventures* (2008–) that has had the biggest impact on American audiences. As a result of these programmes, Karen J. Renner calls ghost hunting 'the new American pastime.'[7] The key difference, as Renner points out, between *Most Haunted* and its American competitor *Ghost Adventures* is the rather gendered framing of the programme. Despite the screaming, and the dodgy evidence, there's something undeniably twee, something very British about *Most Haunted*; it blends a history documentary with a fairly soft approach to the paranormal. Yvette Fielding, as the lead, often chastises her colleagues for startling her or being silly, and demonstrates stereotypically 'feminine' fear. *Ghost Adventures*, on the other hand, is a stag do in a haunted house.

Led by Zak Bagans, a larger-than-life personality, *Ghost Adventures* is less about exploring the history of a place and more about sensational phenomena. If *Most Haunted* is the Oliver Lodge of paranormal TV programmes, *Ghost Adventures* is, without a doubt, Harry Price. Moreover, as we saw in the last chapter, Americans' preoccupation with the paranormal isn't solely in relation to ghosts. *Ghost Adventures* is just as much about demonic forces as it is about the lingering spirits of the dead. This is evident even in the promotional images for the two programmes. Fielding looms, a pleasant or inquisitive expression on her face, transposed over a British castle or surrounded by candles. Bagans, however, looks like he's about to punch a ghost in the face. *Most Haunted* was about exploration, about history; *Ghost Adventures* is, in the vein of Ed and Lorraine Warren, about confrontation, good against evil, the living standing up to paranormal bullies (and, like the Warrens, Bagans has his own museum of haunted artifacts). Bagans is, as Renner describes, 'soldier-like'. There's a sense that the crew of *Ghost Adventures* have to compensate for their fear through mockery and aggression, presenting paranormal phenomena as violent to show that they are suffering from physical attacks rather than jumping at non-physical noises.

As *Most Haunted*'s popularity started to dwindle while *Ghost Adventures* only rose to new heights, there was a sense that they, too, were trying to change their style to keep up with the changing dynamic of paranormal reality television. Yvette's husband Karl and cameraman Stewart began to taunt ghosts ('Come on! Do your worst!' – often to Yvette's expressed annoyance) and, in turn, be viciously attacked. They were hurled down stairs, scratched by phantom claws and hit with projectiles. Later, they added a 'demonologist' to their team in place of a historian, Fred Batt, who would wander around trying to summon Beelzebub into a quaint National Trust property.

Perhaps the biggest direct effect of these international ghost-hunting reality TV programmes is how they inspired their audiences to copy their techniques and try similar investigations for themselves. By 2011, there were 2,500 independent paranormal groups in the UK, compared with just 150 a decade before. This statistic is likely to be much bigger now, fuelled in large part by the proliferation of ghost-hunting videos on YouTube and other video streaming websites and apps. In 2010, YouTube allowed videos on its platform to be up to 15 minutes in length, and steadily, over the years, it increased this to allow constant live-streaming and videos that spanned several hours in length. Along with better and more accessible smartphone cameras and editing software, it meant that paranormal groups could record their own *Most Haunted*-style episodes and garner fame among the ghost-hunting community.

To watch just one of these videos is to forever tweak your YouTube algorithm to endlessly recommend the investigations of other groups and individual ghost hunters. My YouTube homepage is awash with thumbnails of creepy-looking buildings overlaid with the investigator's shocked face. It's really only by doing this that you get a sense of how widespread the phenomenon goes. Moreover, while *Most Haunted*'s influence is undeniably running through every episode purely through its format, this community is very much feeding itself at this point. They follow trends from within the community, such as going to the same locations, and popularise a multitude of equipment that they have seen other teams using. There's more of an air of 1984 cult-classic film franchise *Ghostbusters* about these groups – Bill Murray's character leads a pest-control group of paranormal exterminators, equipped with iconic vacuum backpacks and bleeping electronics. You could choose any random investigation on YouTube and the ghost hunters will most likely have a wide array of electronic equipment with them, including a handful of 'catballs' – cat toys that light up when moved, which are used to allegedly demonstrate when a phantom hand is playing with them. In many ways, these groups have developed their own language full of various strange gadgetry and slang; catball, Rempod, K2, spiritbox, the Estes method. The spiritbox is something akin to Raudive's *Breakthrough* technique which we looked at in Chapter 4, where he believed he could communicate with the spirits through the radio. These little portable radios are designed to rapidly skip through channels in the hope that a ghost will speak through the white noise created. They freak me out, not because I think a ghost will start talking, but the noise they make in general is one of the most awful, maddening things I've ever heard. It's like a thousand people screaming, half a second at a time. Loud, repetitive sounds (car alarms, clocks with an aggressive tick) instantly make me feel like I'm going insane, and I can imagine that, being on a ghost hunt with someone using a spiritbox in front of me, I'd become nervous and jumpy, more inclined to doubt my own scepticism. Perhaps that's the point. Expanding on the spiritbox is what investigators call the 'Estes method', a way of engaging with the device through sensory deprivation. A member of the team wears a blindfold and plugs noise-cancelling headphones into the spiritbox. Other members, unbeknown to the one listening to the spiritbox, ask questions of the ghosts; the member wearing the headphones calls out any words or phrases they hear through the white noise as alleged answers to these questions.

It's rare these days for teams to whack out the old Ouija board. If they do, they surround it with other sensitive equipment designed to detect movement or fluctuations in the local electromagnetic field. But these are no longer as popular as they were during the Satanic panic. They don't necessarily show anything objective unless you're also participating in feeling the glass or planchette move – it doesn't make for good distance viewing. I told someone about my research and he quickly showed me a video on his phone of him struggling to lift a flimsy kitchen chair that he claimed was being pinned down by a ghost; a second later in the video, he suddenly lifted it with ease. He showed it to me with a strange, fearful pride, but I just couldn't feel the second-hand terror he wanted me to – all I saw was him not picking up a chair, and then succeeding in picking up a chair. Ouija boards and any human-centred séance techniques are clearly a had-to-be-there form of spirit communication (apart from the Philip Experiment, which is an uncanny exception). Flashing or beeping electronic gadgets, however, are more impressive to a modern audience, and seemingly eliminate the possibility that a human hand is manipulating them. Often these devices are propped up on a dusty mantelpiece or left in a corner, with the team filming from their phones a few metres away. When the lights of an EMF meter flash yellow or green, it is thrilling – far more thrilling than watching a few fingers on top of a moving glass.

And while the groups do get scared, the investigations are far less reactionary, far less about the presenters themselves, and more about the collection of equipment they bring to each location and capturing these devices 'going off'. As horror critics Sarah Juliet Lauro and Catherine Paul argue, these modern ghost hunters have adopted and appropriated equipment to create am almost mythical interpretation of them: 'Crucial to spiritualist technology, then, is not just the deployment of technological devices, but also an engagement with the methods and rhetoric surrounding them.'[8] This reminds me of what Steve Parsons talked about during the ASSAP training day; these gadgets are just gadgets. If they flash or make a noise, it simply means an LED light or buzzer has been activated. They're not communication devices, but they're used and mythologised as such by these groups.

Even stranger now is the prevalence of phone apps. These claim to detect EMF fluctuations and turn them into other forms of data – words, for instance, or sound. The 'Spirit Talker' app does the latter, shouting out random, vague and spooky words that could be applied to almost any haunted house: 'I died here'; 'drowned'; 'many of us'; '1800s'. The teams then work to apply the words to the information they're aware of, no matter how tenuous – the word 'drowned', for example, might make them suggest that someone died in the river 5km (3 miles) from the property. They work to create evidence out of seemingly coincidental data. While I've seen many groups use these apps, on several occasions they have expressed scepticism for them, wondering how a medieval monk would know how to communicate through a smartphone. Some have tried them once, but never again. This is interesting; technology is the bread and butter of these modern YouTube ghost hunters, but there do seem to be limits to what they feel is acceptable equipment for a spirit to use and be able to interact with.

Another thing that separates online paranormal groups from their terrestrial TV ancestor is the way the majority of locations are abandoned, derelict properties. Particularly in the UK, modern groups are made up of working-class people without the means to spend several hundred pounds hiring out allegedly haunted venues or participating in the ever-popular and ever-lucrative ghost tourism trade. Such locations *are* used – in the UK, especially, groups almost make special pilgrimages to places such as The Ancient Ram Inn in Gloucestershire (indeed, Astrid and Dora from the ASSAP training day had been there the night before) – but social media demands regular content, which puts pressure on groups to visit places without quickly going bankrupt. As such, there's an aspect of urban exploration mixed into the investigation.

But it's not simply down to the fact that YouTube paranormal groups don't have the funds to keep hiring places. There's an aspect of documentation, of remembrance. Yes, everyone wants to see big, explosive poltergeist activity in an abandoned children's home, but a lot of videos involve using paranormal investigation to piece together the history of a building that the rest of society has forgotten. These videos employ the usual techniques – calling out for ghosts, listening for knocking sounds, pointing out a catball flashing in the darkness – but many of them also focus on the things that are left behind when a house is frozen in time. As Michele Hanks argues, there's an aspect of 'political and economic critique' in this practice.[9] In one video, several groups did a crossover investigation into a relatively recently abandoned care home which still contained dozens of pieces of expensive, working equipment. The numerous bedrooms stood empty, but were otherwise in good condition; nature hadn't started to reclaim it yet. The episode became less about coaxing out the spirits of departed residents, and, interestingly, more about the investigators' anger that so much vital equipment was being wasted and so much space was going unused during a time of government austerity, collapsing healthcare and increasing homelessness. Additionally, though, as Hanks describes, it's also a form of protest against being denied entry to sites of cultural and historical importance. Buildings of historical importance are, in the UK, often kept and managed by charitable organisations such as the National Trust, English Heritage, Historic Scotland and Cadw, but there are hundreds of places which have been neglected; today's paranormal investigators try to, in some way, preserve these spaces and make them accessible through their videos.

This practice particularly resonates with me in relation to the letters between Oliver Lodge, Ben Davies and the war-bereaved parents I described in Chapter 6. The address on these letters is a house on Llanbadarn Road. It's a wide, gently winding road that connects Aberystwyth to its neighbouring village of Llanbadarn Fawr, and is populated on either side by elegant, stately 1920s houses. Since I moved in 2022, I drive down this road every time I come home from work. Without fail I imagine, one day, parking my little car on the sweeping driveway of one of these beautiful houses. But when I saw the address on the letters, I couldn't work out to which house it belonged. After a little investigation, I realised that it was the abandoned house that sits, cold and empty and surrounded by construction fencing, not long after you turn onto the road. Call me overly sensitive, but it made me sad – it's a lovely, old, grey-brick house, but I'd always overlooked it in favour of the well-kept houses further along. But through my research, through reading about Mr and Mrs Evans' attempts to speak with the ghost of their son, I feel somehow connected to it, and, just like the local war memorial makes me think of Raymond, I think of Eric every time I pass it. Ghost hunting helps us to remember the places and people who would otherwise fall into obscurity; it helps us to keep the past alive. I obviously don't condone trespassing, but I do sympathise with the way in which modern ghost-hunting groups use their practice to find out the history of a building otherwise left to be forgotten.

While researching this book, I spoke to a number of independent paranormal investigators. Among these was Kris Walsh, part of a group called Paranormal Nerds. Kris and I sat next to each other at the ASSAP training day. Like several other people there, he had plenty of experience in exploring haunted houses but was curious to see the kinds of techniques and training the ASSAP promoted. A former soldier, he struck me as a calm, gentle person and we quickly became friends as the day progressed. I was interested in his process and his group, and rather bombarded him with questions whenever there was a break in the day's programme. Afterwards, we exchanged emails. He told me that his interest in conducting investigations stemmed from a number of childhood experiences which he still cannot explain – and it's almost as though he's retroactively trying to witness those things again now that he has the capability to calmly and logically try to debunk them. Kris retains an element of scepticism in his investigations, calling his approach rational and scientific, but he's come to believe in the paranormal in the sense that he has experienced phenomena he cannot explain. This is what investigations are for Kris – he is open to the strange and uncanny, and once it occurs his task is to find an explanation to the best of his ability before he considers the event to be what he calls 'an unexplained event that is possibly paranormal'.

Kris gave me the run-down of the equipment he uses. Where Eric Dingwall gathered his simple kit of luminous pins, string and chalk, the modern ghost hunter's bag is far more complex. Kris's is all the more fascinating because he builds much of his sensitive equipment himself. He told me that he has a particular interest in this side of ghost hunting, and spends time researching the science of how different sensors work and teaching himself how to build them through YouTube videos (with a few horror movies thrown in for inspiration). By engineering his equipment himself, it gives him a greater sense of control over the devices, as he understands exactly how they work and the circuits that construct them. The Rempod – basically a theremin rod that emits a theremin's creepy screeching sound, accompanied by LED lights, in response to changes in the electromagnetic field – is a particular favourite of Paranormal Nerds, and in addition to the standard rod on a circular base Kris has also built a similar device within a teddy bear that lights up whenever someone (or something) comes near it. I've seen this done by a number of groups. He explained that it's part of their collection of 'trigger objects' which are selected to appeal to a broad range of ghosts. The teddy appeals to the spirits of children, but the group also bring military medals, old keys and coins, and other specific items that he told me are brought out depending on the location's purpose and history.

The technology that Kris uses is actually rather more complex and varied than most groups I've encountered, perhaps because of his impressive skill for constructing and inventing them. Alongside the Rempods, he has made a number of motion detectors and a shadow tracker, and brings along a collection of high-tech sound and video recording devices to ensure that any evidence is captured clearly. And while he also uses several on-trend items among investigators – the light-up catballs, the EMF meter – I was intrigued to learn that the members of his group are still fascinated by more traditional and esoteric forms of attempting to communicate with the spirits. As I mentioned earlier, Ouija boards seem to be falling out of fashion, but for the purposes of experiencing as much phenomena as possible through a variety of techniques, Kris and his group still frequently use them alongside pendulums, bells on a string and divining rods.

Interestingly, however, Kris admitted that recently he's been taking a simpler approach to his investigations. The more he visits haunted locations, the less equipment he uses, stating that for him the only things he really feels are vital are high-quality audio and visual recording equipment. Moreover, he has become increasingly sensitive to the environments themselves, and tailors the approach and the equipment according to the history of the location and the spirits said to still roam there.

In November 2024, Kris invited me back to the Old Prison at Northleach to investigate with his group. While the training day with ASSAP was interesting in its approach to the observation of potentially paranormal phenomena, my research has shown me that ghost hunting has a rich history of experimentation, ad-lib techniques, and is much more sociable than the relative silence in which we first listened out for ghosts in the Old Prison.

I met the group in a local pub for a meal before we set off on our investigation, and this social element was immediately apparent. The majority of them were old friends and had done countless ghost hunts together, but I was instantly welcomed, and everyone was generous in talking to me about their interest in paranormal investigations.

This time, I learnt from my mistake and packed my winter coat, which I've only worn a handful of times when we've had a particularly cold snap. It was a wise move; the Old Prison was very chilly in places, far more so than it was in April, but if any ghosts were making the place even colder I certainly didn't feel it in my toasty duvet coat.

The major difference in investigation styles between this group and ASSAP was, I think, the way in which everyone engaged with the building. As I've previously mentioned, in April we were told simply to stand and observe, but now we were actively appreciating the history and construction of the building. Karin Beasant, a paranormal consultant who regularly works with the Jamaica Inn in Cornwall, was encouraging us to imagine what life would have been like for inmates and workers at the prison. In the courtroom, she even had a few of the gang re-enact a typical (albeit hilariously bawdy) trial. She explained to me that participating in these re-enactments and discussing the history of the building not only allows investigators to connect with the building and its past residents, but may also remind spirits of their life and draw them closer. Despite my scepticism for the latter, I learnt a lot and did feel as though I had a greater understanding of the Old Prison and its history. These investigations are not just about trying to find ghosts; it's another way to empathise with the people who were here before us.

Later in the night, we went inside the former police station attached to the Old Prison. We hadn't had access to this back in April, so Kris was particularly interested to explore this section. It was very cold in there, but seemed somehow modern despite being empty, as though it had until fairly recently been used for offices. We selected a room and various bits of equipment were set up: a camera, a Rempod and several LED devices set to light up in response to atmospheric and temperature fluctuations. As soon as we were settled, Karin asked if I wanted to have a go on her spiritbox using the Estes method. I've mentioned previously that repetitive noises make me feel instantly agitated, but by this point in the evening I was genuinely enjoying myself and I was feeling brave, so I agreed.

Something strange happened, however.

Karin put a blindfold over my eyes, and connected the spiritbox to the headphones I clamped against my ears. I was immersed in juddering static. My heart rate instantly lifted, but I was determined to cope for the sake of the experiment. Occasionally, perhaps every 30 seconds or so, I'd hear the tiniest glimmer of a human voice. I called out what I heard, which wasn't very much, and was so distorted that I think some of it was my brain filling in the blanks (a garbled phrase sounded a little like 'no one can help you now' but equally might have been a snippet of a radio advert for cat food).

I sat like this for perhaps five minutes or so, my shoulders tense and my breathing short as I tried to concentrate despite the maddening sound of the static.

Then the loud, staccato roar suddenly changed. It was no longer a harsh crackle, but a man, a very angry man, bellowing in my ears: 'No! No! No! No!'

My body reacted before my brain, and I violently recoiled, whipping the headphones off. A more seasoned ghost hunter probably would have powered through the visceral fight-or-flight response to listen and analyse the change in sound – I'm sure Mollie Goldney wouldn't have jolted and batted the headphones away in abject horror – but all I knew in that moment was an intense sense of vulnerability, and I had to get away from it.

Once my initial terror had subsided, I had to try to process what had happened. The noise from the spiritbox wasn't recorded, so there was no way to listen back – all I'm left with now is the memory of the sound. My rational explanation is that, after having listened to the static raging in my ears, and feeling increasingly agitated by it, my brain suddenly found a pattern in the chaos and interpreted it as a man's voice. But it wasn't gradual; it was a rapid change, and it no longer sounded remotely like static. It was unlike anything I've ever heard before, and I reacted so strongly because, in that moment, my survival instinct was certain that it was a man I could hear. It wasn't a radio station, either – the snippets I'd previously heard had been no longer than a second.

I could go back and forth like this forever. Each time I grasp at a rational theory, the part of me that felt such intense danger chimes in with a counterpoint. The only thing I can do to get any sort of answer would be to repeat the test – and this, I suppose, is why ghost hunters can never get enough. Each experience demands more experiences in a constant cycle of corroboration.

Far more than during the training day, I truly understood the appeal of ghost hunting that night. Even for a sceptic, it was playful and fun, and in between experiments we drank tea, ate doughnuts and had long conversations about ghosts, horror movies, writing and previous experiences. Throughout the night I felt as though I'd made some lasting friendships with people across the sceptic-believer spectrum. The bonding experience of being in a cold, dark room in the middle of the night, waiting for ghosts to appear, is a peculiarly powerful one. I can't help but think back to all the séances and ghost hunts I've read about since the 1840s, and I wonder how many participants took part simply to spend some time with other people.

We've seen throughout this book that ghosts are incredibly lucrative. This frustration among paranormal investigators about being denied access to haunted locations is something that has been increasingly noticed by heritage properties and private companies. Ghost tourism has a long history in itself.

In 1855, the prominent British writer Harriet Martineau, in *A Complete Guide to the English Lakes*, encouraged the reader to visit Armboth Fell for its haunted house. 'Lights are seen there at night, the people say,' she wrote, 'and the bells ring; and just as the bells all set off ringing, a large dog is seen swimming across the lake. The plates and dishes clatter; and the table is spread by unseen hands. That is the preparation for the ghostly wedding feast of a murdered bride, who comes up from her watery bed in the lake to keep her terrible nuptials. There is really something remarkable, and like witchery, about the house. On a bright moonlit night, the spectator who looks towards it from a distance of two or three miles sees the light reflected from its windows into the lake; and, when a slight fog gives a reddish hue to the light, the whole might easily be taken for an illumination of a great mansion. And this mansion seems to vanish as you approach – being no mansion, but a small house lying in a nook, and overshadowed by a hill.'[10] Martineau's description of a haunted house was part of a general book of travel writing, but it is an early demonstration of how a spooky location can be the source of as much holiday enjoyment as a scenic lake.

More recently, however, and due once again to the influence of *Most Haunted* and the popularity of paranormal investigations as a hobby, ghost tourism is a booming business. These range from the family-friendly to night-long lockdowns tailored to serious investigations. According to Michele Hanks, the first company dedicated to providing ghost tourism experiences opened in 1999, and now most cities offer guided ghost walks as a way to experience the history of a place through its local legends and superstitions. York is a particularly popular destination for these tours, owing largely to its alleged 500 ghosts. These tours, however, are mostly catering to the morbidly curious as an alternative and fun way to experience a location – while still allowing for reflection on historical life and working conditions.

Another form of ghost tourism, that of the private-hire haunted house, seeks to exploit paranormal investigation teams and their constant need to find new locations – many of which in the UK have been featured on *Most Haunted*. One of the most well-known of these is 30 East Drive in Pontefract. Its poltergeist case was similar in some ways to the Enfield haunting; in the late 1960s, the council house became chaotic as the Pritchard family were relentlessly tormented by a violent ghost called 'Mr Nobody' by the media, 'Fred' by the family. Mysterious pools of water appeared throughout the house, sometimes seeping from the walls, 12-year old Diane was dragged up the stairs by her hair and photographs were slashed by kitchen knives. You, too, can be whacked in the face by Fred for £400 per four-person group for a 14-hour experience in the house (the price is for the 'maintenance' of the property; I'll leave you to do the maths on that).[11] For an extra £10, a local medium will come and put a protection spell on you. Theoretical physicists, says the website, can visit for free. The property is now owned by film producer Bill Bungay, who recommends on the house's own website that 'you DO NOT visit, especially if you are of a nervous disposition or have a heart condition'. There's even a waiver that all visitors have to sign, with some interesting clauses. One is that any footage or evidence captured within the house becomes Bungay's copyrighted property, another requests that no Ouija boards are used nor exorcisms attempted so as not to exacerbate the situation. The waiver asks that you don't pinch anything from the house, which is fully furnished in reproduction 1970s style, because Fred might attach himself to the object and follow you home. My favourite clause, though, has to be the one where on signing you agree 'THAT I AM NOT AN IDIOT'. At the ASSAP training day, Steve Parsons had a few choice comments about these sorts of properties – 30 East Drive in particular. As we saw when we looked at poltergeists, cases are violent but fleeting, rarely lasting a few months let alone a few years. Besides, they're most often attached to a person, and the Pritchards have long since moved out. It's remarkable that a poltergeist such as Fred would still be so powerful more than 50 years after he began to make trouble in the property. I think we can safely say that if Harry Price were around today, he would probably have a monopoly on British haunted houses rented out in the name of psychical research.

These sites remain immensely popular among the increasing number of paranormal investigation groups. On the one hand, by privately managing these properties it allows them to be maintained and means that people can experience historical sites that aren't deemed interesting or important enough by official heritage charities. This is more in relation to otherwise derelict old buildings, though, rather than 30 East Drive, which could easily be someone's permanent home. On the other hand, these businesses obviously profit from paranormal investigators needing to visit locations to create content for YouTube or to simply function as a team. With 30 East Drive's fame being helped by *Most Haunted*, it's become a place that seems to validate the existence of some paranormal investigators – you're not legitimate unless you've been. I would like to see ghost tourism meet somewhere in the middle. If struggling heritage properties leaned into people's natural curiosity over ghosts and ghost hunting, they might see a surge in visitors; if private businesses made their haunted properties more financially accessible to everyone, it would help to preserve the history of these places and allow people to conduct investigations on a larger scale. I'm doubtful, though.

Much of this chapter has examined the ways technology and electronic gadgets now pervade the modern ghost-hunter's kit. Far removed from the chalk and string of Eric Dingwall's box of equipment, our search for life after death is now wholly through recorded data and the often mythical ways in which we interpret what we capture on phones or through EMF meters. In the final chapter, as I come to a conclusion about why we keep searching for ghosts, we'll look at what the future of paranormal investigations might look like, and whether our preoccupation with using technology might ultimately draw us away from searching for ghosts at all.

Conclusion

As we saw in the previous chapter, technology is now a vital part of the experience of ghost hunting. Paranormal investigators are rarely without their EMF meters, creating narratives and personalities of spirits through flashing lights. Throughout this book, we've seen how changing technologies have equally changed the rhetoric of spiritual survival after death. Messages from the dead were received like a telegram, then ectoplasmically materialised on photographs and spoke through the static of the wireless. Now, the 'energy' of spirits manipulates electromagnetic fields, messing with recording equipment and draining the batteries of phones and torches to become powerful enough to produce phenomena.

Urban legends surround technology, but where once it was used as a metaphor to explain spirit communication, it seems that increasingly technology is seen as something that ghosts can manipulate and inhabit. In March 2024, Amazon collaborated with Yvette Fielding of *Most Haunted* fame, asking customers to send in videos of paranormal phenomena captured on its Ring CCTV cameras.[1] From the entries, she identified a clip of a storage unit as being the spookiest.[2] The clip shows a large space with a car visible in the greyscale night vision. A peculiar doughnut-shaped 'orb' floats lazily, randomly, around the room. Its movements are not that of dust or an insect. I watched it a few times, thinking it was perhaps the circle of light from a torch as a potential burglar scoped out the premises – but the shape doesn't change, whereas a torch beam shifts and moves around objects. It was peculiar. Elsewhere on the internet, viral videos crop up of Tesla cars being driven through cemeteries. On its dashboard screen, it shows that the car's sensor is detecting human shapes where there aren't any.

We seem to view technology as something inherently possessable – something that ghosts can manipulate and control regardless of whether they lived 400 years before the invention of the iPhone. Jeffrey Sconce describes our persistent habit, despite the proliferation and normalisation of technology in our everyday lives, of ascribing 'mystical powers' to our gadgets and devices.[3] It is as Eleanor Sidgwick explained when she was exposing spirit photography. We don't fully understand the process of how these things work, and therefore leave ample room for folklore and superstition to creep into the circuit boards. Moreover, we're always recording, always capturing – from security doorbells to car dashcams to video game live-streams. With this abundance of material, it's only to be expected that recorded evidence of anomalous phenomena increases too.

But will there be a point where technology becomes so advanced that it permanently alters our perception of life and death, and the very nature of bereavement? Spiritualism was at its peak following the First World War, when the grieving public sought to communicate with the young men whose lives were violently cut short on the battlefield. It ultimately, however briefly, changed the way the public grieved. As we saw in the letters exchanged between Ben Davies and Oliver Lodge, it was commonplace and perfectly socially acceptable for the bereaved to cope with their loss through seeking a medium to connect them to the recently deceased. Since then, our relationship with the paranormal has become increasingly reliant on technology to stand in place of the medium. A ghost is now defined by the lights on a cat toy or the words that spew forth from an app, and the future of ghosts may become problematic because of this.

If ghosts are becoming increasingly 'materialised' through technology, when will the line become so blurred between data and paranormal activity that technology *creates* its own spirit? Something I became more and more aware of during the course of writing this book was the rise of artificial intelligence – specifically, the use of AI as a conversation partner. My students tell me they use AI on a variety of apps to help with their mental health, either venting to it or simply as a way to have an uncannily realistic chat with someone. Within this ever-advancing technology are apps and websites that allow users to tailor their AI chat-partner to follow particular speech and personality rules and to act like a particular person.

Largely, these bots are used in a fictional context, but an increasing number are based on real people. If you wanted, you could discuss the final season of *Game of Thrones* with an AI based on Leonardo da Vinci. Renowned figures from history talk again through AI, but, to some extent, this is still within the realm of fiction – particularly figures from ancient history. It's similar to the test Professor Anderson conducted in Chapter 1, when he had seven different mediums in seven different séances simultaneously produce the spirit of George Washington. We saw how the ghosts of famous people seemed to become public property during the height of Spiritualism, and the same seems to be true of these AI bots. However, people are now beginning to create private chatbots of their recently departed loved ones, tailoring them to sound exactly like the writing style of the deceased. Tech startups are recognising our innate, timeless need to communicate with the dead, and AI chatbot engines are being designed specifically for this purpose with somewhat ghoulish names like Here After AI and Seance AI.[4] It seems as though AI is becoming a modern Ouija board, its generated messages working as something like automatic writing. Indeed, automatic writing was believed to be produced from the unconscious mind, detached enough that it felt alien to the writer. AI works in the same way – it all comes from a repository of training, prompted in the right way, and feels like a genuine, spontaneous reaction.

There were times in this book when I laughed at some of the things I wrote about, but equally there were moments that genuinely upset me, when my ability to take a rational and objective approach was complicated by the palpable grief I uncovered in my research. I couldn't dismiss Sir Oliver Lodge's conversations with Raymond because of how deeply it helped him and his wife deal with their overwhelming sadness. I found W. T. Stead's alleged messages from the Blue Island ludicrous, but again my ability to criticise wavered in the face of empathising with what his daughter must have gone through. For every case of exploitation of the bereaved, I've read another example of how genuinely soothed people have been by the thought of speaking with their deceased loved ones. When we looked at Summerland, I briefly examined Margaret Cameron's *The Seven Purposes*, which described a number of exchanges between the bereaved and spirits residing in the utopian Spiritualist afterlife. At one point, Cameron explained how she had been helping her friend speak to her dead son over a number of séances. The grieving woman changes for the better as a result: 'The next day Mrs Gaylord went home, where she immediately destroyed all her black-bordered cards and stationery and similar symbols of mourning. She wrote me that she felt it was false and wicked to mourn for a son as vitally alive and happy as she now knew Frederick to be.'[5]

There have been many occasions while writing this book when my hands stopped typing and I sighed, needing a few minutes to untangle how I really felt about certain moments in the history of ghost hunting. When I first began, I think I treated this project as a bit of tongue-in-cheek fun. I bought Bev the Haunted Doll because I thought it was hilarious. I printed out and stuck my ASSAP Accredited Investigator certificate to my office door, knowing it would amuse people. I wasn't expecting to be so moved and, at times, incredibly troubled by the cases and testimonies I later read about. I no longer think that ghost hunting is a fool's errand. That's not to say I now believe in ghosts, but I don't see the practice as a waste of time for the superstitious and the gullible. Far from it. I've come to realise that ghost hunting isn't really about finding ghosts at all. It's about so much more than that; it's so much more *complicated* than that. It's about offering solidarity and empathy when someone else is frightened; it's about connecting with history, with old buildings, with forgotten lives; it's about having fun with like-minded people; it's about interrogating our own curiosity, our own scepticism; it's about negotiating ourselves with our own fears and hopes about death.

I asked Kris about what he thought ghost hunting might look like in the near and distant future. His answer was fascinating. He told me that, recently, he's seen a much more amicable relationship between what he called 'technologists' and 'traditionalists' – modern paranormal investigator lingo for Spiritualists and scientists. Thinking back, this described the group at the ASSAP training day. We all worked together on the investigation regardless of our approach to ghost hunting. Dora had been 'sensitive' to spirits since childhood, and despite my scepticism we got on well as we sat in the dark on the top floor of the Old Prison. This is how paranormal groups operate now: all are welcome. They are a mix of sceptics and believers, bringing their own experiences, and their own equipment, and they take a broader, more tolerant, more sociable approach to ghost hunting. Kris said this will only continue to improve in the next few years. It's not blurring the boundary between sceptics and believers, who are still firmly opposed to the other's ideas on what ghosts are, but it creates a friendlier environment for both sides to work together without the heated and, quite frankly, violent debates of the last two centuries.

He told me that technology would increasingly be used in investigations but, crucially, it would involve 'more of the same'. This simple phrase represents so much of what I've seen about the history of ghost hunting; it seems like it has changed dramatically through YouTube culture and use of fancy gadgets, but really the process hasn't changed at all since the Fox sisters. An interesting part of his group's process, he said, was to sit down and chat casually, and to try not to consciously think about what it was they were there to do. If they started to hear knocks or felt the temperature change, they ignored it for a particular length of time while they continued to talk about light, friendly, non-spooky topics. I told him that this is exactly what Spiritualist circles did over 150 years ago, but with more hymns, and he was surprised – it was just something they'd come up with to create a warm and welcoming atmosphere in dark, spooky environments. No matter how technology is employed, the process remains fundamentally the same: a group of friends, listening out for anything they could consider to be paranormal phenomena.

Dr Kate Cherrell, whom I spoke to about First World War Spiritualism, echoed this shift in trends to do things more 'authentically' with a return to Victorian Ouija boards, planchette writing and table-tipping, which she argues is a 'reactionary' move against the fraudulent practices of sensational paranormal television programmes. Paranormal investigations on the whole, though, are becoming more social than ever owing to their popularity on video streaming websites and apps. For every video posted, there are dozens of comments from viewers helping to spot evidence that the group may have missed. This is, however, leading to the downfall of paranormal programming on television, which can no longer compete with the apparent raw, amateur authenticity of streaming websites. Nevertheless, Kate believes that paranormal investigations will always captivate and intrigue us, and will therefore always be a lucrative business.

Steve Parsons, the ASSAP's training officer, is a little more cynical about the future of paranormal investigations. Amateur ghost-hunting groups are the bane of his existence as he perseveres in trying to organise investigations into following a code of best practice, and while he agreed with Kate Cherrell that the bubble of paranormal reality TV shows has burst, YouTube remains an increasingly popular form on which these independent groups operate. He thinks things will become more muddled, more ethically dodgy, and more in keeping with some of the more peculiar Spiritualists we've met throughout this book – citing that many groups now offer exorcisms and cleansing, and that investigations are, in line with Zak Bagans, about battling good and evil spirits. Just as Kris mentioned that he's increasingly keeping his investigations simple, Steve predicts that the popularity of ghost-hunting tech is merely a phase which is already on its way out. He told me that 'people are waking up to the fact that many of the devices which they have invested significant amounts in simply don't work' and that the growing number of devices being listed on eBay shows that these groups are trying to recoup the costs of the technology on which they've spent a considerable amount of money.

For Steve, there will always be reports of paranormal phenomena and there will always be dedicated teams who follow a strict code of professional conduct to investigate cases. But he thinks that, largely, the commercialisation of ghost hunting sparked by *Most Haunted* (of which he is vociferously not a fan) has had its day. Ghost tours will likely start to dwindle in popularity, and privately owned 'haunted' locations will soon struggle to attract the numbers of live-streaming paranormal groups in quite the same way. But ghost hunting will never cease to fascinate us; as he told me, 'the public will always love to be scared.'

I think that ghost hunters will continue to adapt to new technologies, to use and appropriate them into methods of evidence-capture and communication, but it will be more of the same. My PhD student, Kieran Judge, is currently working on the puzzle of what ghost stories will look like in the far future. What happens when people begin to die in space, on the Moon, on Mars? We spend our meetings contemplating haunted lunar bases and poltergeists on space stations. And, ultimately, we come to the same conclusion every time, the same prediction as Kris: more of the same. The haunted house may no longer be made of crumbling brick in the future, but its effect on our imagination will be the same, and we'll continue to investigate it with the same approach as psychical researchers in the late nineteenth century. As long as death is certain, our scepticism will remain thoroughly uncertain.

I didn't encounter a ghost while writing this book. In the end, I don't think I ever really expected to. But while I've always considered myself to be sceptical, all the things I've read, the places I've been and the people I've spoken to have made me reflect on what scepticism and belief means beyond myself. I feel like, in some ways, belief is communal whereas scepticism is personal. Perhaps this is why Spiritualists grouped together in circles and churches, whereas their fiercest critics so often worked alone. Despite some of the ludicrous cases I've come across which have made me laugh, I think I've learnt how personally significant it is for some people to believe in ghosts. The turning point was, I think, Lady Lodge's comment: 'We can face Christmas now.' Time and again, I've read about how communication with spirits has allowed people to work through their grief, to make them better able to face the day without their loved one at their side. Yes, paranormal investigations and Spiritualism are rife with fraud, manipulation, exploitation, but I've encountered far too many people whose hearts were mended by their belief in the utopian Summerland for me to still feel comfortable in dismissing other people's experiences. At the beginning of this project, I thought the research would be a bit of silly fun; I wasn't expecting it to be so morally complex.

I've come to reconsider what my own scepticism stands for, too; while I didn't experience anything I could concretely call paranormal, I found that my disbelief is shakier than I thought it would be. In the dark of the Old Prison during my paranormal investigator training, the moment I shared with Dora still intrigues me. And I think of the way I stared into Bev's plastic blue eyes as the light in my spare room audibly fizzled above me. I realised how badly I wanted something more, something tangible, something completely inexplicable, to happen to me.

While I started writing this book, I was trying to find a ghost. Actually, though, I don't this book was about ghosts at all; it's about grief. Grief lingers in every chapter, every anecdote, every experiment, whether people were conscious of it at all. This is what I've come to conclude; ghosts manifest from grief. I don't necessarily mean immediate bereavement, either, nor do I mean the death of a loved one. Ghosts are our defence against the general dread and aftermath of death. This book has been about two sides of a debate: believers and sceptics. But I think there's another dualism that has been at war throughout this story: those who accept death and those who reject it. Sceptics and Spiritualists alike may, from their opposing corners, both work to better come to terms with what happens after we die. And I say 'we' because I think, after all, that paranormal investigations are fundamentally about ourselves, the living, rather than the dead. Andrew Jackson Davis, gazing upon the utopia of Summerland, accepted his fate. We saw early on in this book how Harry Houdini wanted nothing more than to speak to his late mother; time and again he was disappointed. I think Houdini rejected death. That is why we investigate ghosts, because we know that mortal death is inescapable – the ultimate loss of control in our neat and ordered lives – and we search not just for proof but for a way to make peace with that part of ourselves, that ghost within us, which almost violently cannot comprehend finality.

So, no, I did not find a ghost while writing this book. But I think, now, when the midnight existential dread creeps upon me, I understand those feelings better. I understand better my own mortality, my own constant grief for people I've lost and will lose in the future, and I understand that my own scepticism is fragile because, ultimately, I'm someone who stares into the void and needs to see something stare back.

I'll keep looking.

Acknowledgements

This book benefitted from many fascinating conversations with paranormal investigators, researchers and parapsychologists, and I want to thank Dr Kate Cherrell, Steve Parsons, Bill Eyre, Kris Walsh, Anna Goodrich, John Hasted, and everyone at the ASSAP AAI training day for taking the time to talk to me about their work and involvement with ghost hunting. I'm also very grateful to Tansy Barton and the staff at Senate House Library for their help and advice in consulting their archive.

My editors at Bloomsbury Sigma, Jim Martin and Sarah Lambert, have been so much fun to work with. I'd also like to express my gratitude, as always, to Donald Winchester at Watson, Little for his unwavering enthusiasm for my weird ideas. And, last but not least, thank you for the incredible illustrations, Cassie. I still can't believe we got to work together! I hope you've recovered from *Ghostwatch*.

References

Introduction

1 Shepherd, R. M. 2009. 'Dangerous consumptions beyond the grave: Psychic hotline addiction for the lonely hearts and grieving souls', *Addiction Research and Theory*, 17 (3), 278–290, p. 278.
2 Musella, D. P. 2005. 'Gallup poll shows that Americans' belief in the paranormal persists', *The Skeptical Inquirer*, 29 (5), p. 5.
3 French, C. C., Haque, U., Bunton-Stasyshyn, R. *et al.* 2009. 'The "Haunt" project: an attempt to build a "haunted" room by manipulating complex electromagnetic fields and infrasound', *Cortex*, 45, pp. 619–629.
4 Hill, J. A. 1918. *Man is a spirit: a collection of spontaneous cases of dream, vision and ecstasy*. New York: George H. Doran Company.
5 Schmitt, J-C. 1998. *Ghosts in the Middle Ages: The Living and the Dead in Medieval Society*, translated by Teresa Lavender. Chicago; London: The University of Chicago Press, pp. 2–3.
6 Ferber, S. 2004. *Demonic Possession and Exorcism*. New York: Routledge, p. 24.
7 *Ibid.*, p. 25.

Chapter 1

1 O'Donnell, E. 1955. *Haunted People*. London: Rider and Company, p. 54.
2 Morton, L. 2015. Ghosts: A Haunted History. London: Reaktion Books, Ltd., p. 66.
3 Owen, A. 2004 [1989]. *The Darkened Room: Women, Power and Spiritualism in Late Victorian England*. Chicago, London: The University of Chicago Press, p. 50.
4 Britten, E. H. 1868. 'Rules to be Observed for the Spirit Circle', *Human Nature*, pp. 48–52.

5 Stowe, H. B. 1867. 'The Other World', *Religious Poems*, Boston: Ticknor and Fields.
6 McGarry, M. 2008. *Ghosts of Futures Past: Spiritualism and the Cultural Politics of Nineteenth-Century America*. Berkeley: University of California Press, pp. 29–30.
7 Luxmoore, J. C. 1873. 'Gross Outrage at the Spirit Circle', *Spiritualist*, 12 December.
8 Volckman, W. 1874. 'My Ghost Experience: The Struggling "Ghost"', *Medium and Daybreak*, 16 January.
9 Crookes, W. 1874. *Researches in the Phenomena of Spiritualism*. London: J. Burns, p. 3.
10 Anderson, J. H. 1853. *A Shilling's Worth of Magic, or, Tricks to be Learnt in a Train*. London: R. S. Francis.
11 Houdini, H. 1924. *A Magician Among the Spirits*. New York; London: Harper & Brothers, p. xii.
12 Owen, I. M. & Sparrow, M. 1976. *Conjuring Up Philip: An Adventure in Psychokinesis*. Toronto: Fitzhenry & Whiteside, p. xv.

Chapter 2

1 Crookes, W. 1874. *Researches in the Phenomena of Spiritualism*. London: J. Burn, p. 16.
2 Underwood, P. 2000. *The Ghost Club Society*. Surrey: White House Publications, p. 3.
3 Myers, F. W. H., Gurney, E. & Podmore, F. 1886 [1866]. *Phantasms of the Living*, Vol. 1. London: Trübner and Co., p. xxxvi.
4 *Proceedings of the Society for Psychical Research*, Vol. 1. 1883. London: Trübner and Co., p. 4.
5 Pearsall, R. 1972. *The Table-Rappers*. London: Michael Joseph, p. 68.
6 Sidgwick, E. 1887. 'Results of a Personal Investigation into the Physical Phenomena of Spiritualism: with Some Critical Remarks on the Evidence for the Genuineness of Such Phenomena', *Proceedings of the Society for Psychical Research*, Vol. 4. London: Trübner and Co., pp. 45–74.
7 Ben Davies papers. 1894–1903. Located at: National Library of Wales, Aberystwyth. BOX 17/8.

8 Pennington, Rev. M. 2007. 'Deliver us from evil…', *The Churches Fellowship for Psychical and Spiritual Studies*, https://tinyurl.com/5xxkeare [accessed 26 May 2024].
9 Parsons, S. T. 2018. *Guidance Notes for Investigators of Spontaneous Cases*, New Edition. London: Society for Psychical Research, p. 1.

Chapter 3

1 Morton, R. C. 1892. 'Record of a Haunted House', *Proceedings of the Society for Psychical Research*, Vol. 8. London: Kegan Paul, Trench, Trübner and Co., Limited, pp. 311–332.
2 Owen, I. M. & Sparrow, M. 1976. *Conjuring Up Philip: An Adventure in Psychokinesis*. Toronto: Fitzhenry & Whiteside, p. 88.
3 Flammarion, C. 1924. *Haunted Houses*. New York: D. Appleton and Company, p. 74.
4 Price, H. 1940. *The Most Haunted House in England*. Bath: Cedric Chivers, Ltd., p. 1.
5 Sargent, E. 1887. *Planchette; or, the Despair of Science*. Boston: Roberts Brothers, p. 1.
6 Price, P. 1946. *The End of Borley Rectory*. Bath: Chivers Press, p. 69.
7 Scrapbook BE. c.1896–1959. Located at: Senate House Library, London. MS192/1/27 93.
8 Scrapbook BE. c.1896–1959. Located at: Senate House Library, London. M3912/1/27 94.
9 Scrapbook BE. c.1896–1959. Located at: Senate House Library, London. MS912/1/27 106A.
10 Dingwall, E. J., Goldney, K. M. & Hall, T. H. 1956. *The Haunting of Borley Rectory*. London: Gerald Duckworth & Co. Ltd., p. 72.
11 Borley Rectory Scrapbook. c.1956–c.1957. Located at: Senate House Library, London. MS912/1/38/2 7.
12 Collins, B. A. 1948. *The Cheltenham Ghost*. London: Psychic Press Limited.
13 Underwood, P. 1994. *Nights in Haunted Houses*. London: Headline, pp. 9–10.

14 Braithwaite, J. P., Perez-Aquino, K. & Townsend, M. 2004. 'In search of magnetic anomalies with haunt-type experiences: pulses and patterns in dual time-synchronized measurements', *The Journal of Parapsychology*, 68 (2), 255–288, p. 271.

Chapter 4

1 Nicholson, A. & Blakey, A. 2021. 'Ghost hunters claim to capture *The Shining* twins on camera at "UK's most haunted hotel"', *Manchester Evening News*, 24 June, https://tinyurl.com/y78u8ewm [accessed 17 July 2024].
2 Kaplan, L. 2008. The Strange Case of William Mumler, *Spirit Photographer*. Minneapolis: University of Minnesota Press, pp. 28–29.
3 Renaud, G. 'Prix et salaires à Paris en 1870 et 1872', *Journal de la société statistique de Paris*, 14 (1873), 176–185, p. 177.
4 Krauss, R. H. 1995. *Beyond Light and Shadow*. Munich: Nazraeli Press, pp. 129–130.
5 Sidgwick, E. 1892. 'On Spirit Photographs: A Reply to Mr A. R. Wallace', *Proceedings of the Society for Psychical Research*. London: Kegan Paul, Trench, Trübner and Co., Limited, p. 276.
6 Coates, J. 1911. *Photographing the Invisible: Practical Studies in Spirit Photography, Spirit Portraiture, and other Rare but Allied Phenomena*. London: L. N. Fowler & Co., pp. 61–62.
7 von Schrenck-Notzing, A. 1923. *Phenomena of Materialisation: A Contribution to the Investigation of Mediumistic Teleplastics*, translated by Fournier d'Albe, E. E. London: Kegan Paul, Trench, Trübner and Co. Ltd., p. 1.
8 Crawford, W. J. 1921. *The Psychic Structures at the Goligher Circle*. New York: E. P. Dutton & Company, p. 1.
9 Raudive, K. 1971. *Breakthrough: An Amazing Experiment in Electronic Communication with the Dead*, translated by Fowler N. Gerards Cross: Colin Smythe, p. 2.
10 Nees, M. A. & Phillips, C. 2015. 'Auditory Pareidolia: Effects of Contextual Priming on Perceptions of Purportedly Paranormal and Ambiguous Auditory Stimuli', *Applied Cognitive Psychology*, 29, pp. 129–134.

11 Konstantin Raudive – Breakthrough (Full Recording), YouTube, 12 November 2017, https://youtube/bxRTguQtHfw?si=-WVi4UKaWujVnhRV [accessed 18 March 2024].

12 Chalfen, R. 2008. '*Shinrei Shashin:* Photographs of Ghosts in Japanese Snapshots', *Photography & Culture*, Vol. 1 (1), pp. 51–72.

Chapter 5

1 Davis, A. J. 1868. *A Stellar Key to the Summer Land*. Boston: William White & Company, p. 21.

2 Davis, A. J. 1878. *Views of Our Heavenly Home: A Sequel to A Stellar Key to the Summer-Land*. Boston: Colby & Rich, p. 20.

3 Carr, K. M. *c*.1939. *Spirit-World Teachings*, ed. by Shirley Carson-Jenney. London: Arthur H. Stockwell Limited, p. 20.

4 Davis, A. J. 1973. *The Diakka, and Their Earthly Victims; Being an Explanation of Much that is False and Repulsive in Spiritualism*. New York: A. J. Davies & Co., p. 7.

5 Stead, W. T. 1909. *How I Know the Dead Return*. Boston: The Ball Publishing Co., p. 6.

6 Stead, W. T. 1922. recorded by Woodman P. and Stead E. *The Blue Island: Experiences of a New Arrival Beyond the Veil*. London: Hutchinson & Co., p. xiii.

7 Ward, J. S. M. 1920. *A Subaltern in Spirit Land*. London: William Rider & Son, Limited., p. 12.

8 Cameron, M. 1918. *The Seven Purposes: An Experience in Psychic Phenomena*. London: Harper & Brothers Publishers, p. 195.

9 *Golden Gate*, Vol. 9 (6), 24 August 1889.

10 'Observe Fortieth Anniversary of Summerland Church', *Spiritualist Monthly*, April 1932, p. 25.

11 Summerland tourist page. Visit Santa Barbara, https://tinyurl.com/2wr73wnw [accessed 29 July 2024].

12 Koudounaris, P. 2023. 'The Rainbow Bridge: The True Story Behind History's Most Influential Piece of Animal Mourning Literature', *The Order of the Good Death*, 9 February, https://tinyurl.com/4jfrnt5v [accessed 5 August 2024].

Chapter 6

1 Lodge, O. 1916. *Raymond, or Life and Death: with examples of the evidence for survival of memory and affection after death*. New York: George H. Doran Company, p. 17.

2 White, S. E. 1946. *The Stars Are Still There*. New York: The Golden Eagle Press, p. 65.

3 Kelway-Bamber, L. 1919. *Claude's Book*. New York: Henry Holt and Company, p. xvi.

4 Barker, E. 1915. *War Letters from the Living Dead Man*. New York: Mitchell Kennerley, pp. 6–7.

5 Durand, G. G. 1917. *Sir Oliver Lodge* Is *Right*. Illinois: privately printed, p. 17.

6 Patrick, R. 2013. 'Speaking across the borderline: Intimate connections, grief and spiritualism in the letters of Elizabeth Stewart during the first world war', *History Australia*, 10 (3), 109–128, p. 124

7 Mercier, C. A. 1917. *Spiritualism and Sir Oliver Lodge*. London: The Mental Culture Enterprise, p. v.

8 Winter, J. 1995. *Sites of memory, sites of mourning: The Great War in European cultural history*. Cambridge: Cambridge University Press, p. 58.

9 'Police Officer as Major: Sensational Evidence', *Cambrian News and Merionethshire Standard*, 3 May 1918, p. 5.

10 Hazelgrove, J. 2000. *Spiritualism and British Society Between the Wars*. Manchester: Manchester University Press, p. 17.

11 Gildart, K. 2018. 'Séance Sitters, Ghost Hunters, Spiritualists, and Theosophists: Esoteric Belief and Practice in the British Parliamentary Labour Party, c. 1929–51', *Twentieth Century British History*, 29 (3), 357–387, p. 366.

12 Ellwood, R. S. 1993. *Islands of the Dawn: The Story of Alternative Spirituality in New Zealand*. Honolulu: The University of Hawaii Press, p. 52.

13 Ben Davies Papers. 1882–1950. Located at: National Library of Wales, Aberystwyth. BOX 3/1–16 Letter 614.

14 Ben Davies Papers. 1882–1950. Located at: National Library of Wales, Aberystwyth. BOX 3/1–6 Letter 872.

15 Ben Davies Papers. 1882–1950. Located at: National Library of Wales, Aberystwyth. BOX 3/1–6 Letter 873.

16 Ben Davies Papers. 1943. Located at: National Library of Wales, Aberystwyth. BOX 17/12.

Chapter 7

1 Bering, J. M., Smith, S., Stojanov, A., *et al.* 2022. 'The "Ghost" in the Lab: Believers' and Non-Believers' Implicit Responses to an Alleged Apparition', *The International Journal for the Psychology of Religion*, 32 (3), pp. 214–231.
2 Price, H. 1925. *Stella C.: An Account of Some Original Experiments in Psychical Research*, London: Hurst & Blackett Ltd.
3 Price, H. 1931. *Regurgitation and the Duncan Mediumship.* London: National Laboratory of Psychical Research.
4 From Mrs K. M. Goldney. Located at: Senate House Library, London. HPC/4B/84.
5 Ghost Hunting Kit. Located at: Senate House Library, London. HPG/1/9/1.
6 Fuller, J. G. 1979. *The Airmen Who Would Not Die.* London: Souvenir Press, p. 34.

Chapter 8

1 Owen, A. R. G. 1964. *Can We Explain the Poltergeist?* New York: Helix Press, p. 133.
2 Thurston, H. 1953. *Ghosts and Poltergeists.* London: Burns Oates, p. 80.
3 Grottendieck, W. G. 1906. 'A Poltergeist Case', *Journal of the Society for Psychical Research.* London: The Society's Rooms, 260–266, p. 261.
4 French, C. 2024. *The Science of Weird Shit: Why Our Minds Conjure the Paranormal.* Cambridge, MA: MIT Press, p. 73.
5 'The Sauchie Poltergeist' Report (BX). 1961–1965. Located at: Senate House Library, London. MS912/1/43.
6 Green, A. 2016. *Ghost Hunting: A Practical Guide, the new edition*, ed. Murdie, A. Suffolk: Arima Publishing, p. 87.
7 Playfair, G. L. 1981. *This House is Haunted: The Investigation of the Enfield Poltergeist.* London: Sphere Books, Ltd., pp. 1–2.

8 Sitwell, S. 1940. *Poltergeists: An Introduction and Examination Followed by Chosen Instances*. London: Faber and Faber Limited, p. 82.
9 Fodor, N. 1959. *The Haunted Mind: A Psychoanalyst Looks at the Supernatural*. New York: Garrett Publications, p. 72.
10 Fodor, N. 1958. *On the Trail of the Poltergeist*. London: Arco Publications, p. 31.
11 'Which of These Two Fodors is Right?' *Psychic News*, 8 October 1938.

Chapter 9

1 Howitt, W. 1865. 'The Devils of Morzine', *Cornhill Magazine*, Vol. 11, 468–481, p. 469.
2 Britten, E. H. 1883. *Nineteenth Century Miracles; or, Spirits and Their Work in Every Country of the Earth*. Manchester: William Britten, pp. 380–381.
3 Bartholomew, R. B. & Hassall P. 2015. *A Colourful History of Popular Delusions*. New York: Prometheus, p. 127.
4 Harris, R. 1997. 'Possession on the Borders: The "Mal de Morzine" in Nineteenth Century France', *Journal of Modern History*, 69 (3), 451–478, p. 476.
5 Mark 5:1–20. King James Version.
6 Warren, L., Warren, E., with Chase, R. D. 1990. *Ghost Hunters: True stories from the world's most famous demonologists*. London: Futura, p. 3.
7 Roberts, E. 1959. *Forty Years a Medium*. London: Herbert Jenkins, p. 32.

Chapter 10

1 *Most Haunted*, Series 1 Episode 1, 'Athelhampton Hall', Antix Productions, 2002.
2 Manning, L. dir. *Ghostwatch*. BBC, 1992.
3 Besterman, T. ed. 1934. *Inquiry into the Unknown: A B.B.C. Symposium*. London: Methuen & Co.
4 Haunted House Broadcast. 1936. Located at: Senate House Library, London. HPE/1/2.

5 Ofcom, *Ofcom Broadcasting Code*, https://www.ofcom.org.uk/siteassets/resources/documents/tv-radio-and-on-demand/broadcast-codes/legacy-codes/broadcast-code-2005.pdf?v=331614 [accessed 10 October 2024].
6 Ofcom, *Ofcom broadcast bulletin* 49, 5 December 2005, p. 16.
7 Renner, K. J. 2013. 'Negotiations of Masculinity in American ghost-hunting reality television', *Horror Studies*, 4 (2), pp. 201–219.
8 Lauro, S. J. & Paul, C. 2013. '"Make Me Believe!" Ghost-hunting technology and the postmodern fantastic', *Horror Studies*, 4 (2), 221–239, p. 226.
9 Hanks, M. 2016. *Haunted Heritage: The Cultural Politics of Ghost Tourism, Populism, and the Past.* Oxon: Routledge, p. 165.
10 Martineau, H. 1855. *A Complete Guide to the English Lakes.* London: Whittaker and Co., p. 70.
11 30 East Drive, http://www.30eastdrive.com/booking-step-1/ [accessed 1 August 2024].

Conclusion

1 Ring, 'The search for Ring's most haunted', 18 March 2024, https://tinyurl.com/4v8kh49t [accessed 19 August 2024].
2 Tinney, A. 2024. 'Phantom-like orb caught in shocking Ring Doorbell footage after "paranormal activity"', *Daily Star*, 29 April, https://tinyurl.com/4hnu9475 [accessed August 19 2024].
3 Sconce, J. 2000. *Haunted Media: Electronic Presence from Telegraphy to Television*. Durham; London: Duke University Press, p. 6.
4 Agarwal, M. 2023. 'The race to optimize grief', *Vox*, 21 November, https://tinyurl.com/4fun4hy4 [accessed 19 August 2024].
5 Cameron, M. 1918. *The Seven Purposes: An Experience in Psychic Phenomena*. London: Harper & Brothers Publishers, p. 58.

Index

Acorah, Derek 242–243, 249–250
Adelphi Hotel, Liverpool 91–92
aeroplanes 181–182
afterlife, Spiritualist 117–137, 139–140, 163
air raids 233–234
airships 182, 185–189
aliens 121–123
Alvey, Helen 161–162
American Society for Psychical Research 43
Anderson, John Henry 28–31, 34
animals, in Spiritualist afterlife 135–136
Arthaud, Dr 220
artificial intelligence (AI) 271–272
Association for the Scientific Study of Anomalous Phenomena (ASSAP) 49–55, 56, 58–62
Athelhampton Hall 241–244
Athenodorus 3–4
automatic writing 128–133, 135–136, 145–156, 206, 272

Bagans, Zak 253, 275
Balfour, Arthur 41, 163
Barbanell, Maurice 229
Barham, HMS 176
Barker, Elsa 150–151
Barrett, Sir William 41, 43
Bartholomew, Robert E. 222
Batt, Fred 254
BBC 70, 82, 206, 242, 244–245
Beadnell, Charles Marsh 108
Beasant, Karin 262, 263
Beenham, Ethel 187
Bell, Alexander Graham 43
Bennett, Sir Ernest 244
birds, in Spiritualist afterlife 136
Bisson, Juliette 99, 101, 102–103
The Blue Island: Experiences of a New Arrival Beyond the Veil 129–133
Bolacre, Gilles 193
Bond, Elijah 234
Borderland 125
Borley Rectory, Essex 70–84, 87, 232–234, 244
Breakthrough 108–115
British College of Psychic Science 182, 187
British National Association of Spiritualists 41
Broadcast of a Haunted House, The (radio programme) 245
Brockway, Almira 159
Brown, Thomas 87
Brown, William 172–173
Buguet, Édouard 95–96, 97
Bull, Reverend Henry 71–72
Bungay, Bill 266

Cambrian News 158
Cambridge Association for Spiritual Inquiry 40
Cameron, Margaret 136–137, 272–273

Campbell, Virginia 191–193, 198–200, 205–206, 209
Carrière, Eva 98–103
Carrington, Hereward 108
Castlefield haunted house, Manchester 49–50, 51
Catholicism 6–7, 218, 223
celebrity spirits 29–30, 110–112
Charles, Craig 245, 247
Chase, Robert David 225
chatbots 271–272
Cherrell, Kate 143–144, 274–275
Christian Science churches 138
Christianity 5–7, 55–56, 117–118, 138, 223
Churches Fellowship for Psychical and Spiritual Studies (CFPSS) 55–56
Churchill, Winston 111
clairvoyance 180–189
Claude's Book 148–150
Clyne, Edna 139
Coates, James 97
College of Psychic Studies 229
Colley, Thomas 46
Collins, B. Abdy 84–86
A Colourful History of Popular Delusions 222
The Conjuring film franchise 208, 224–225, 231, 233, 248
Constans, Dr 220–221
Cook, Florence 19–27, 44, 124, 171
Cooke, Andrea 241–242, 243
Cooke, Patrick 241–242
Cooper, Mr 232–234
council houses 200–208, 215, 266–267
Cranshaw, Stella 167–168
Crawford, William Jackson 104–108, 180
Crookes, Sir William 19, 23–27, 39–40, 44, 69
cryptozoology 49
cults 137–139

Daily Mirror 70, 203–205
Dale, Arthur 13
Dante Alighieri 134
Darwin, Charles 117–118, 139–140
Davenport, Ira 32–33
Davenport, William 32
Davies, Benjamin 47–49, 161–164, 258
Davis, Andrew Jackson 119–124, 136, 137
Day, Horace 16
demonic forces 6–7, 124, 217–227
Despard, Rose 85–86
Diakka 123–124
The Diakka, and Their Earthly Victims 123–124
Dingwall, Eric 80–84, 108, 172, 177–178, 180
dinosaurs 135
direct voice mediums 128, 227–231
dirigibles 182, 185–189
disintegrating forces 136–137
Divine Comedy 134
dolls, possessed 231, 235–240
Dostoevsky, Fyodor 111
Doyle, Sir Arthur Conan 27, 40, 42, 157, 184, 185, 187, 188
Dr Constans 222
Duncan, Helen 168–177, 212, 227
Durand, Grace Garrett 151–152
Durham, Bishop of 40

Earl, Mrs 182–184
ectoplasm/teleplasm 93, 100, 101–102, 104, 105–106, 156, 168–176, 227
Edmonds, Judge 95
Eglinton, William 46–47
Egyptian archaeology 69
electromagnetic field (EMF) meters 57–58, 256
electromagnetic fields (EMF) 2–3, 257
The End of Borley Rectory 80, 232
Enfield poltergeist, London 202, 203–208, 209, 210, 225
equipment *see* technology/ equipment
Estes method 255, 263–264
Evans, Eric 162
Evans, Ivor 162–164
evolution, theory of 117–118, 139–140
exorcism 6–7, 219, 222, 223, 224, 227
Exorcist, The (film) 224
expectation 37, 68
extra-terrestrial life 121–123
Eyre, Bill 49, 51

Ferber, Sarah 6–7
Ferdinand, Archduke Franz 150–151
Fielding, Yvette 242, 243, 249, 253–254, 270
First World War 128, 133, 140, 141–161
Flammarion, Camille 69–70, 119, 121, 156
flying brick photograph 80, 83
Fodor, Nandor 209–214
Forbes, Mrs 210–214
Fowler, Nadia 114
Fox, Kate 15, 16, 28–29, 33, 44
Fox, Margaret 15, 28–29, 33
Freemantle, Francis 160
French, Chris 2–3, 198
Freud, Sigmund 34, 43, 213, 214
fuji séances 15
Fuller, John G. 182

galvanic batteries 30
Gardiner, Mike 205
Garrett, Eileen 180–189
Gef the Talking Mongoose, Isle of Man 209
Ghost Adventures (TV programme) 113, 252–253
Ghost Club 40–41, 47, 125
Ghost Club Society 86–87
Ghost Hunters 225
Ghost Hunting: A Practical Guide 201
ghost-hunting societies *see* psychical research societies
'The Ghost in the Stereoscope' photographs 93–94
The Ghost Society 40
ghost tourism 257, 264–267, 275–276
Ghostbusters film franchise 254–258
Ghostwatch (TV programme) 244, 245–247
Glanville, Helen 78–79
Golden Gate 137
Goldney, Mollie 77, 80–84, 169, 171–173, 174, 176–177
Goligher, Kathleen 104–108
Green, Andrew 201
Greene, Sarah 245–246, 247
Grisewood, Mr 245
Grosse, Maurice 204–205, 210
Grottendieck, W. G. 194–196
Guidance Notes for Investigators of Spontaneous Cases 56–58

Guppy, Elisabeth 94
Gurney, Edmund 41, 45

Hall, Radclyffe 145
Hall, Trevor 80–84
Hall, Trevor C. 192
Hanks, Michele 258, 265
Hardinge Britten, Emma 16–17, 36, 94–95, 97, 221
Harris, Ruth 222
Hassall, Peter 222
Hasted, John 197
haunted buildings 3–4, 49–56, 63–89, 230–231
  Borley Rectory, Essex 70–84, 87, 232–234, 244
  Castlefield, Manchester 49–50, 51
  ghost tourism 257, 264–267, 275–276
  haunted school 50–55
  investigator training 58–62
  Muncaster Castle, Cumbria 88–89
  Old Prison, Northleach 58, 60–62, 261–264, 273–274
  parapsychology 87–89
  'Record of a Haunted House' 64–69, 84–86
  'Stone Tape Theory' 69–70
  see also poltergeists
Haunted Houses 69–70
The Haunting of Borley Rectory 80–84
Heeps, Carolyn 203
Henning, Reverend A. C. 74–75
Hinchcliffe, Emilie 183, 184, 185
Hinchcliffe, W. G. R. 181–182, 183–184, 185
Hitler, Adolf 111, 112
Hodgson, Janet 202, 203–208
Home, Daniel Dunglas 20, 23–24, 39–40
Houdini, Harry 27, 31–34, 156
'How I Know the Dead Return' 126, 131
'How the Mail Steamer Went Down in Mid Atlantic, by a Survivor' 125–126
Howitt, William 218, 219, 221–222
Hudson, Frederick 94, 95

Immram stories 131
independent paranormal investigation groups 57, 60–61, 254–264, 267, 273–276
Inquiry into the Unknown (radio programme) 244
International Institute for Psychical Research 209, 210–211, 213
Irving, James 209
Irving, Voirrey 209

James, Henry 43
James, M. R. 40
James, William 43
Jenney, Shirley Carson 123, 135–136
Jesus 223
Joan of Arc 151
Johannesburg poltergeist 194, 196
Johnson, Alice 44
Judge, Kieran 276
Jung, Carl 111
Jupiter 122
Jürgenson, Friedrich 109

Karl, Jason 243
Kaye, Lucie 71
Kelly, William 160
Kelway-Bamber, Claude 147–150

Kelway-Bamber, L. 147–150
King, Katie 19–27, 44, 124, 171
Kneale, Nigel 70
Krauss, Rolf H. 96

laboratory experiments 165–166
  investigations of mediums 167–177, 180–189, 227
  paranormal voice recording 108–115
Labour Party 160
Lauro, Sarah Juliet 256
LBC Radio 205
Leckie, Malcolm 157
Lenin, Vladimir 111
Leonard, Gladys 145–146, 147, 148, 150, 153–154, 164
levitation 20, 27, 208, 210
Lincoln, Abraham 95, 151
Lincoln, Mary Todd 95
lithobolia 193–196
Loder-Symonds, Lily 157
Lodge, Alec 153–154
Lodge, Lady 154–155, 161, 276
Lodge, Lionel 153–154
Lodge, Raymond 142–146, 152, 153–154, 161
Lodge, Sir Oliver 47–48, 69, 97, 111, 142–146, 147, 148–149, 151, 152, 153–156, 160, 161–162, 244, 258
Logan, Sheila 198–199
Logan, William 198–199
London Spiritualist Alliance 148, 163, 182, 184, 185
London Stereoscopic Company 93–94

McBain, Cindy 225–227
McGarry, Molly 18
McKenzie, James Hewat 182
A Magician Among the Spirits 31–32
magicians 27–34, 46, 156
magnetic field anomalies 88–89
Martineau, Harriet 265
materialisation mediums 19–27, 44–45, 46, 93, 98–108, 124, 168–177, 210–211, 227
Maurina, Zenta 109
May, Madam 158–159
mediums 13–34
  automatic writing 128–133, 135–136, 145–156
  clairvoyance 180–189
  direct voice mediums 128, 227–231
  ectoplasm/teleplasm 93, 100, 101–102, 104, 105–106, 156, 168–176, 227
  during First World War 142–161
  and haunted buildings 50, 54, 55
  laboratory investigations of 167–177, 180–189, 227
  laws against 158–160, 164, 176–177
  magicians' investigations of 27–34, 46, 156
  materialisation mediums 19–27, 44–45, 46, 93, 98–108, 124, 168–177, 210–211, 227
  planchette séances 78–79, 183
  psychometry 228–231
  scientific investigations of 23–27, 39–40, 43–49
  during Second World War 163–164
  slate-writing mediums 33, 45–47
  spirit photography 91–108, 115–116
  'spirit rescues' 54, 55
  spirits' descriptions of Summerland 124–133, 135–137
Mellon, Annie Fairlamb 44–45

Mercier, Charles A. 155–158
Moon 123
Morris, Graham 208
Morton, R. C. 64–69, 84–86, 180
Morzine, French Alps 217–222
The Most Haunted House in England 75, 81–82, 179
Most Haunted (TV programme) 60, 241–244, 247–254, 265, 266, 267, 276
Motte, Pierre de la 6–7
Mumler, William 94–95, 97
Mumler, William H. 44
Muncaster Castle, Cumbria 88–89
Myers, F. W. H. 65
Myers, Frederick 41, 45
Myers, Frederick William Henry 145

National Association of Spiritualist Churches of New Zealand 160
National Laboratory of Psychical Research 168–177, 180–189
Nawrocki, G. B. 79
Nees, Michael A. 112–113
New England Society for Psychical Research 224
Nights in Haunted Houses 87
Nottage, George Swan 93
Nottingham, Peggy 203, 204

objects, possessed 231–240, 270
Obry, Nicole 6–7
O'Donnell, Elliott 13
Ofcom 250–252
Ofn (TV programme) 251
O'Keeffe, Ciarán 250, 251
Old Prison, Northleach 58, 60–62, 261–264, 273–274
On the Origin of Species 117–118, 139–140
Onesiphorus, Sir George 60
'The Other World' (poem) 17–18
Ouija boards 76, 78–79, 183, 224, 225–226, 234–235, 239, 240, 255–256, 261, 274
Owen, A. R. G. 34, 192–193, 198, 199, 206
Owen, Iris 68

Palladino, Eusapia 33, 98, 102
Paranormal Nerds 259–264
paranormal voice recording 108–115
parapsychology 139, 164, 189
  haunted building investigations 87–89
  poltergeist investigations 197–198, 199, 205, 206–214
pareidolia 113–114
Parkinson, Michael 244, 245–247
Parsons, Steve 56–58, 59, 199, 256, 266, 275–276
Patrick, Rachel 153
Paul, Catherine 256
Phantasms of the Living 41
Phenomena of Materialisation 98–103
'Philip Experiment' 34–37, 68, 167–168, 256
Phillimore, Mercy 163–164
Phillips, Charlotte 112–113
phone apps 257
Photographing the Invisible 97
photography, spirit 91–108, 115–116
Plagnat, Julienne 218–219
Plagnat, Marie 218
planchette séances 78–79, 183

Playfair, Guy Lyon 202, 203, 206–208, 210
Pliny the Younger 3
Podmore, Frank 41
poltergeists 189, 191–215
  and children 191–209, 215
  and council houses 200–208, 215, 266–267
  Enfield poltergeist 202, 203–208, 209, 210, 225
  Johannesburg poltergeist 194, 196
  Pontefract poltergeist 266–267
  Sauchie poltergeist 191–193, 198–200, 205–206, 209
  stone throwing 193–196
  Sumatra poltergeist 194–196
  Thornton Heath poltergeist 209–214
Pontefract poltergeist 266–267
Price, Harry 70–84, 166–177, 178–180, 181, 184, 187–189, 204, 209, 212, 232–234, 244, 245
Proceedings of the Society for Psychical Research 42
  'Record of a Haunted House' 64–69, 84–86
Protestantism 5–6
Psychic News 213–214
The Psychic Structures at the Goligher Circle 104–108
psychical research societies 39–62
  Association for the Scientific Study of Anomalous Phenomena (ASSAP) 49–55, 56, 58–62
  Ghost Club 40–41, 47, 125
  Ghost Club Society 86–87
  International Institute for Psychical Research 209, 210–211, 213
  investigator guidance and training 56–62
  Society for Psychical Research (SPR) 41–49, 56–58, 63–69, 70, 84–86, 96, 145, 163, 166–167, 194–196, 204–205
  Toronto Society of Psychical Research 34–37, 68, 167–168, 192
psychokinesis 197–198, 199, 201, 208, 209
psychometry 228–231

radio programmes 205, 244–245
Raudive, Konstantin 108–115
Raymond, or Life and Death 143–146, 147, 151, 153–156
reality television programmes 241–254, 274–276
  Ghost Adventures 113, 252–253
  Ghostwatch 244, 245–247
  Most Haunted 60, 241–244, 247–254, 265, 266, 267, 276
  Ofn 251
'Record of a Haunted House' 64–69, 84–86
Red Cloud ('spirit guide') 228, 230–231
Regurgitation and the Duncan Mediumship 212
reincarnation 148–149
Rempods 255, 260, 263
Renner, Karen J. 252, 253
Rickard, Bob 49
Roberts, Estelle 227–231
Rogers, Edmund 41
'Rules to be Observed for the Spirit-Circle' 16–17, 36
Ruskin, John 43

Satanic panic 224–227, 234, 239
Saturn 119, 122

Sauchie poltergeist, Scotland 191–193, 198–200, 205–206, 209
Scherman, Mr 80
Schmitt, Jean-Claude 5
school, haunted 50–55
Schrenck-Notzing, Albert von 98–103, 187
Sconce, Jeffrey 270
séances 13–37, 118
  automatic writing 128–133, 135–136, 145–156
  clairvoyance 180–189
  direct voice mediums 128, 227–231
  ectoplasm/teleplasm 93, 100, 101–102, 104, 105–106, 156, 168–176, 227
  during First World War 142–161
  fuji séances 15
  laboratory investigations of 167–177, 180–189, 227
  laws against 158–160, 164, 176–177
  magicians' investigations of 27–34, 46, 156
  materialisation mediums 19–27, 44–45, 46, 93, 98–108, 124, 168–177, 210–211, 227
  'Philip Experiment' 34–37, 68, 167–168, 256
  planchette séances 78–79, 183
  psychometry 228–231
  scientific investigations of 23–27, 39–40, 43–49
  during Second World War 163–164
  sexualised nature of 103, 107–108
  slate-writing mediums 33, 45–47
  spirit photography 91–108, 115–116
  spirits' descriptions of Summerland 124–133, 135–137
Second World War 161–164
The Seven Purposes: An Experience in Psychic Phenomena 136–137, 272–273
Shelley, Percy Bysshe 135–136
shinrei shashin photographs 115
Sidgwick, Eleanor 41, 43–47, 96–97, 270
Sidgwick, Henry 41, 42, 45
Sir Oliver Lodge Is Right 151–152
Sitwell, Sacheverell 209
Slade, Henry 33, 45–46
slate-writing mediums 33, 45–47
Smith, Reverend G. E. 71, 74
Snowden, Ethel 160
Society for Psychical Research (SPR) 41–49, 56–58, 63–69, 70, 84–86, 96, 145, 163, 166–167, 194–196, 204–205
Sparrow, Margaret 68
spirit possession 215, 217–240
  demonic forces 6–7, 124, 217–227
  direct voice mediums 128, 227–231
  exorcism 6–7, 219, 222, 223, 224, 227
  Morzine, French Alps 217–222
  possessed objects 231–240, 270
'spirit rescues' 54, 55
'Spirit Talker' app 257
Spirit-World Teachings 123, 135–136
spiritboxes 255, 263–264
Spiritualism 13–34
  automatic writing 128–133, 135–136, 145–156, 206, 272

clairvoyance 180–189
cults 137–139
direct voice mediums 128, 227–231
ectoplasm/teleplasm 93, 100, 101–102, 104, 105–106, 156, 168–176, 227
during First World War 128, 133, 140, 141–161
laboratory investigations of mediums 167–177, 180–189, 227
laws against mediums 158–160, 164, 176–177
magicians' investigations of 27–34, 46, 156
materialisation mediums 19–27, 44–45, 46, 93, 98–108, 124, 168–177, 210–211, 227
planchette séances 78–79, 183
psychometry 228–231
scientific investigations of 23–27, 39–40, 43–49
during Second World War 161–164
slate-writing mediums 33, 45–47
spirit photography 91–108, 115–116
'spirit rescues' 54, 55
Summerland 117–137, 139–140, 163
and Thornton Heath poltergeist case 213–214
see also spirit possession
Spiritualism and Sir Oliver Lodge 155–158
Spiritualist Success Church 138
The Stars are Still There 146–147
Stead, Estelle 127–133
Stead, W. T. 40, 125–133, 136
Stead, Willie 127, 130
Steer, Lee 91–92
Steer, Linzi 91–92
A Stellar Key to the Summer Land 119–120
Stewart, Elizabeth 153, 161
Stewart, George 153
Stewart, Miss 192, 198, 206
The Stone Tape 70
'Stone Tape Theory' 69–70
stone throwing 193–196
Stowe, Harriet Beecher 17–18
Strutt, John William 42–43
A Subaltern in Spirit Land 133–135, 136
Sumatra poltergeist 194–196
Summerland 117–137, 139–140, 163
Summerland commune, California 137–139
Swift, Jonathan 123
Swinhoe, Henry 85
Swinhoe, Imogen 85–86

Tabori, Paul 83–84
Tavernier, Joseph 218, 219
Tavernier, Peronne 218, 219, 222
technology/equipment 92, 267–268, 269–272
artificial intelligence (AI) 271–272
chatbots 271–272
electric-shock device 48–49
electromagnetic field (EMF) meters 57–58, 256
galvanic batteries 30
ghost-hunting equipment 57–58, 77, 177–180, 254–258, 260–261, 263–264, 269, 274–275, 276
laboratory equipment 167, 169, 173, 184

Ouija boards 76, 78–79, 183, 224, 225–226, 234–235, 239, 240, 255–256, 261, 274
paranormal voice recording 108–115
phone apps 257
Rempods 255, 260, 263
spirit photography 91–108, 115–116
spirit possession of technology 270
spiritboxes 255, 263–264
video streaming 10, 56, 252, 254–258, 260, 267, 274, 275
telekinetic energy 189, 197, 199, 209
television programmes *see* reality television programmes
Thatcher, Margaret 201
Thomas, Daniel 158–159
Thornton Heath poltergeist, London 209–214
Thurston, Luke 197
Titanic 125, 127–133
Tolstoy, Leo 111, 151
Toronto Society of Psychical Research 34–37, 68, 167–168, 192
tourism, ghost 257, 264–267, 275–276
Tours poltergeist, France 193
training, for paranormal investigators 58–62
Troubridge, Una 145
Turner, Cathy 87
Turner, James 87

UFOs 49, 59
unconscious mind 34, 109, 272
Underwood, Peter 40, 41, 86–87
Uvani ('spirit guide') 180, 183–184, 185–186, 187, 188

vaginal examinations 169, 172–173
Vagrancy Act of 1824 158, 159, 164, 176
Venus 123
video streaming 10, 56, 252, 254–258, 260, 267, 274, 275
Views of Our Heavenly Home 119, 120–122
Volckman, William 20–22, 124, 171

Wall, V. C. 70
Walsh, Kris 259–263, 273–274
War Letters from the Living Dead Man 150–151
Ward, J. S. M. 133–135, 136
Warren, Ed 208, 224–227, 231
Warren, Lorraine 208, 224–227, 231
wartime grief and Spiritualism
First World War 128, 133, 140, 141–161
Second World War 161–164
Washington, George 29–30
Wells, H. G. 121
Westcott, Reverend B. F. 40
White, Stewart Edward 146–147
Winter, Jay 157
Witchcraft Act of 1735 159, 160, 176
Wood, C. E. 44–45
Woodman, Pardoe 127–133
Wriedt, Etta 128

X-ray examinations 173–174, 212

Yeats, W. B. 40, 101
YouTube 10, 56, 114, 206, 252, 254–258, 260, 267, 274, 275

Zeppelins *see* dirigibles